PRAISE FOR THIS BOOK

"For the Roman Catholic, there is no higher nor more powerful expression of worship than the Latin Mass. Lieutenant Colonel David Sonnier knew this and spent his exemplary military career attempting to bring this beautiful liturgy to the troops he served alongside. With a deep and bitter spiritual war currently engulfing our world, this book is a critical retrospective into the lives of those service members who had already discovered the key to victory. Lt. Col. Sonnier shows us how true victory is not simply an overwhelming of the enemy's defenses resulting in ceded territory; the true and final measure of the victorious man is his faithfulness in all trials while leaving the outcome to God. If you are a Catholic service member who loves the beauty of family life and the transcendence of traditional worship, Sonnier's book is a must-read!"

—ROBERT A. GREEN JR., Commander, USN, author of *Defending the Constitution Behind Enemy Lines: A Story of Hope for Those Who Love Liberty*

"Catholics who serve in the military are accustomed to making personal sacrifices but they should never be made to sacrifice their access to Holy Mass in its traditional form. This book is the inspiring account of a career soldier in the Army who strove tirelessly to make the Latin Mass accessible to Catholics in uniform. His campaign for the Mass faced institutional obstacles and unjust biases but his battle stride never slackened. This story will resonate with every Catholic, civilian and military, who is committed to the Holy Faith and will inspire fellow Catholics to fight the good fight. As a parish priest and retired military chaplain, I applaud this courageous soldier of Christ!"

—FR. JOHN PAUL ECHERT, Lieutenant Colonel, USAF (ret.)

"Lt. Col. David Sonnier's masterful story of the struggle to recover and maintain Roman Catholic tradition in the US Military provides crystal-clear insights into the current ecclesiastical malaise, its root causes, and achievable solutions. The military is a remarkable microcosm of the entire country; this analysis offers detailed explanations of conflicts and tensions familiar to Catholics in nearly every diocese in America. Sonnier pivots between historical-theological concepts and relatable personal vignettes that draw the reader in. *Rites and Wrongs* speaks directly to crises in military culture but goes beyond that to encompass broader issues of church, society,

and geopolitics. The author names names and places, meticulously attributing all his sources. This book throws you into the 'foxhole' with a professional Army officer who put everything on the line to win opportunities for future Catholic military members to pray the Mass of All Ages at their assigned camp, post, or station."
—JOHN W. PROCTOR, Sergeant Major, US Army (ret.)

"The chaplaincy is among the oldest institutions in American military life. The U.S. Army Chaplain Corps traces its origins to July 29, 1775, when Congress passed a resolution establishing pay for chaplains within the Continental Army, officially recognizing the clergy already serving within colonial militias. George Washington himself took the matter seriously, writing that chaplains should be men of 'character and good conversation' who would 'influence the manner of the corps both by precept and influence.'

"In *Rites and Wrongs*, Lt. Col. David Sonnier shows how the guidance of our first general has not always been followed. With dramatic detail, he describes his and others' efforts to obtain the Latin Mass for military members, in accordance with the pastoral guidelines of John Paul II. He reveals the broader pressure within the chaplain corps toward a vaguely therapeutic, ecumenically diluted model of ministry—one that treats the specifically religious, sacramental function of a priest as optional rather than central.

"Those who love the Traditional Latin Mass and serve in the U.S. military have never asked for special treatment. The Archdiocese of the Military Services, caught between institutional inertia, post-Vatican II ecclesiastical politics, and a genuine priest shortage, has so far declined to open the doors to the TLM, which prevents young, committed orthodox priests from serving our military members. *Rites and Wrongs* adds much to the current discussion and is a well-documented institutional and personal history. As a Naval Aviator flying in eight squadrons, I always privately included prayer in my takeoff checklist. Our military members are much more spiritually minded than people realize and deserve the choice to attend the TLM in chapels, in the field, and at sea."
—MATTHEW BRASMER, Commander, USN (ret.)

Rites and Wrongs

Rites and WRONGS

One Man's Struggle for the Latin Mass in the U.S. Army

LT. COL. DAVID SONNIER

Os Justi Press

Os Justi Press
P.O. Box 21814
Lincoln, NE 68542
www.osjustipress.com

Send inquiries to
info@osjustipress.com

ISBN 978-1-965303-99-3 (paperback)
ISBN 979-8-90378-000-6 (hardcover)
ISBN 979-8-90378-001-3 (ebook)

Book design by Michael Schrauzer
Cover photograph by Amy B. Proctor,
used with permission

For the laity throughout the world who, after *Summorum Pontificum*, rose to the occasion to put to rest the myth that "nobody is interested in that anymore."

For all priests who in any way assisted in lifting the restrictions on the traditional Latin Mass — especially those who learned the Latin Mass themselves or helped establish a community dedicated to Catholic tradition.

For those bishops who accepted the explanation from Pope Benedict XVI, modified their understanding of the situation, and took active steps to make the Latin Mass available — and especially bishops who, subsequent to *Traditionis Custodes*, continue to protect and make provision for those who are dedicated to Catholic tradition.

CONTENTS

PREFACE

THE EVENTS DESCRIBED IN THIS BOOK OCCURRED AT a time during which there were no rules governing the requirements for the existence of the traditional Latin Mass within a diocese. In 1988 Pope John Paul II had simply asked bishops, kindly, please, to respect the "rightful aspirations" of those who desired to worship according to the old rite. A few years later, the question of who had these "rightful aspirations" and how they could make them known was still lingering. The 1988 guidance on the traditional Latin Mass had come from Pope John Paul II in *Ecclesia Dei*, written in response to the illicit consecrations of four bishops by Archbishop Marcel Lefebvre, founder of the Society of St. Pius X (SSPX). It was met with minimal cooperation from bishops. However, as a general rule, if the SSPX happened to have a presence within the diocese, the bishop would make a nearby provision for an Indult Latin Mass. As a result, a Latin Mass that had the approval of the diocese was typically only available when there was an SSPX apostolate in close proximity. Without the presence of an SSPX chapel, the bishop had absolutely no incentive to make any such provision. This lack of clarity on the status of the traditional Latin Mass, and the fact that it was completely unavailable across great swaths of the United States or anywhere within military chapels, set the stage for the events described in this book.

After leaving the U.S. Army and moving on to a career in academia, I told very few people about these events. People I did tell either did not believe it — including family members and even fellow Catholics who attended the traditional Latin Mass — or they would dismiss it with a comment along the lines of "nobody is interested in that anymore." I believed at that time that it was important to document the events, and I'm grateful to Goretti Publications for publishing the first edition of this book. When it was published in 2007 under the title *Rightful Aspirations*, the traditional Latin Mass was still not even on the radar of most Catholics. Prior to the publication of Pope Benedict's motu proprio *Summorum Pontificum* later that same year, the Latin Mass was hidden from the public, poorly understood, and accessible to very few. It was something widely misunderstood as dangerous, schismatic, or problematic for reasons nobody could explain. The events described in this book were already becoming a distant memory to all involved. By the year 2007 the U.S. was deeply involved in conflicts in Iraq and Afghanistan, so the previous edition and the subject it addressed were unlikely to receive much attention.

Summorum Pontificum demonstrated that the notion that "nobody is interested in that anymore" was false. As it turned out, the restrictions on the Latin Mass had been kept in place for long enough to propagate a myth that there was no interest. Once those restrictions were lifted, a new reality quickly emerged. Not only was there considerable interest, but that interest grew and blossomed into a movement that threatened to bring about the collapse of an empire built on widespread erroneous notions about Vatican II and even the nature of the Church Herself. *Traditionis Custodes* was a hastily conceived and poorly thought-out response to that growing movement. It attempts to sweep the problem under the rug instead of addressing it. The problem that must eventually be addressed is that we are living with an understanding of Vatican II which was founded on lies that were propagated in the aftermath of the council.

Astute Catholics have known for some time that we were in for a final wave of persecution, and we are seeing that as this edition goes to publication. Latin Masses are being cancelled around the U.S. as a dying ideology based on propaganda, lies, and hatred for Catholic tradition thrashes about in its death throes. This book is written with sympathy for those who are undergoing spiritual suffering as a result. I sincerely hope that hearing a story similar to theirs will comfort them, ease their suffering, and help them to see a way forward. By way of *Traditionis Custodes*, those bishops charged with the care of souls were temporarily granted the freedom to inflict cruel measures on the faithful. Some of these aging prelates seem incapable of understanding that young people yearn for Catholic tradition. Many young people, including recent converts, embrace the richness of Catholic tradition with vigor. It is heartbreaking to see a young family move to the vicinity of a Latin Mass only to have it eliminated by some elderly bishop dreaming of a return to the 1970s. To those young families, I extend my heartfelt compassion for what they are going through.

The spiritual care of those who have chosen to serve in the U.S. Armed Forces is of utmost importance, and these souls have been neglected. Petition drives are practically non-existent among those serving in uniform, so they are unable to speak on their own behalf. In the doctrinal confusion and chaos of the post-Vatican II era, the decline in Mass attendance was accompanied by a corresponding decline in the Catholic understanding of marriage, abortion, the Eucharist, and other fundamental truths of our faith. It is no surprise that the moral principles of a "Just War" have also been ignored with the same casual disinterest, creating a situation that is dangerous for the entire world. To allow our powerful military to lose its moral restrictions on the use of force is foolish and dangerous. The widespread presence of tradition-minded Catholics among all branches of the armed forces should be encouraged as quickly as possible. As Cardinal Sarah recently said in an interview:

> How we believe is how we pray. If we don't have faith, we can't take action. If people don't believe, nothing will change. We continue to fight over the liturgy; we continue to bully certain people. Whereas in fact, when we really look at the Christians who practice today, they are the ones who go to traditional Mass. So why forbid them? On the contrary, we should encourage them.[1]

One can argue that it is just one cardinal's opinion that the people who attend the traditional Mass are the ones who practice their faith regularly. It is helpful to note that his observation corresponds to a 2020 study by Fr. Donald Kloster,[2] and it corresponds very well to the reality seen by anyone paying attention. At some point this reality must be recognized. We have a large, dangerous military that should be as full as possible of people of all ranks who have a traditional Catholic belief in the value of human life and an understanding of the conditions under which the use of military force is justified. The best way to ensure that this becomes the state of our armed forces as quickly as possible would be to work to restore traditional Catholic practice and belief among our servicemen and women.

There is nothing stopping us from doing that.

---

[1] "Cardinal Sarah in interview: I have spoken to the Pope about the Traditional Mass, he is aware, he is the Father of all," *Rorate Caeli*, October 14, 2025, https://rorate-caeli.blogspot.com/2025/10/cardinal-sarah-in-interview-i-have.html.

[2] "The Latin Mass Among Millennials & Gen Z: A National Study," *The Missive*, June 5, 2020, https://fssp.com/latin-mass-among-millennials-study/.

ACKNOWLEDGMENTS

I WRITE THIS WITH ETERNAL GRATITUDE TO MY WIFE, who stood beside me during this entire episode and through-out the years that have followed. Lorri's countless virtues as a Catholic wife become obvious to anyone following this saga.

I must thank each of my children, as they have all contrib-uted to this effort in some way. The boys dutifully learned to serve the Mass, embraced the traditional Catholic teachings, and follow in our footsteps in their own way, despite the fact that we're often quite different from those around us. Several of them now wear the uniform, and long conversations with them in recent years helped me to understand the need for this revised edition. Many thanks to my oldest daughter, Sr. Mary Clair, MICM, for the suggestion that I update the 2007 edition of this book (*Rightful Aspirations*, Goretti Publications) and my younger daughter, Annie Inferrera, for her insistence on strict adherence to grammar rules and punctuation precision.

Heidi Akridge, an excellent editor with a history of working for various Army publications, proved to be exactly the right person for the final editing and formatting. Her experience as an Army wife and a mother enabled her to make countless helpful suggestions regarding content.

All of us owe a debt of gratitude to Dr. Peter Kwasniewski, the publisher, whose intense work on behalf of the Church is unparalleled in recent history. When I first looked up from my busy life and started reading his excellent essays, articles, and books, my first thought was that this is an American version of Michael Davies. Since then I've become convinced that his con-tributions are even more consequential. The clarity of thinking that he provides is essential during this time of confusion. It is my constant prayer that American Catholics will be the means by which the Church is restored. If this prayer is eventually answered, it will certainly be through the efforts of people like Dr. Kwasniewski.

Several priests contributed greatly to the recommendations found in the last chapter of this book. I am most grateful to Fr. John Echert, who serves as a chaplain with the Air National

Guard, and Fr. Kenneth Bolin, West Point class of '96 and former U.S. Army chaplain. Both priests directed me to the applicable laws, statutes, and canons governing the Archdiocese for the Military Services. Fr. Charles Johnson, an active-duty chaplain at Parris Island who routinely celebrates the traditional Latin Mass, assisted greatly with my understanding of the current state of the military chaplaincy and the trends within recent years. I spoke with several other priests who chose not to be mentioned, but who nevertheless helped me to understand the current situation so as to be able to make recommendations that are realistic. Retired Sergeant Major John Proctor also assisted by suggesting steps that can be taken for an immediate improvement in the lives of Catholics serving in the armed forces. These suggestions and recommendations are all found in the final chapter of this book. I am grateful to Amy Proctor for allowing us to use, for the cover, her striking photograph of a chaplain offering the Mass of the Ages at a military base.

Finally, I must thank the many officers and enlisted men I worked with who assisted in the effort described in this book. In particular, I must thank Retired Brigadier General Earnie Callender, U.S. Air Force, whose career was rudely terminated because of the assistance he provided to tradition-minded Catholics whose only desire was to be treated as nicely as the Pagans and Wiccans were being treated by the U.S. Army Chaplains Corps. May God reward him, and may history vindicate him.

Rites and WRONGS

Chapter 1

THE END OF AN ERA

O N A HOT AUGUST AFTERNOON IN 1997 I WAS standing in the office of the senior Fort Bragg chaplain, a Colonel and a Catholic priest. By all expectations our meeting should have been cordial. I have always had the highest respect for priests, and the two of us seemingly had everything in common. We were both serving in the U.S. Army, we were both Catholic, and both of us were field grade officers assigned to Airborne units at Fort Bragg. It should have been a pleasant encounter, but instead he was shrieking at me like a raving lunatic. He was frothing at the mouth and spit was flying. A vein in his forehead was popping out. He wiped his profusely sweating brow.

"I just want to talk about our request," I said.

"There's nothing to talk about!" he shouted back.

"But Cardinal Ratzinger..."

"I don't answer to him! Go to him! Don't come to me!" Spittle was flying.

I stared at him, speechless, waiting for some inspired words to come to me, or maybe time for him to gather his wits. He quit panting, sat down behind his desk and pretended to be reading a report. After a few seconds he looked up:

"This meeting is over. Finished! Get out of here!" he barked.

I didn't leave. He stood again.

"Get out of here!" Now he was screaming. "Get out of my office! You piss me off!"

I still didn't leave.

"You are dismissed, major! This meeting is over! Now get out! Go!"

I just stood there until he reached for the phone. "Listen, Major, you get the hell out of my office now or I'm calling the military police!"

I turned slowly and left, walking absentmindedly out the door and back to my pickup truck.

Maybe I should have stayed and allowed myself to get arrested. That would have caused a stir! The headlines would scream *"Green Beret Arrested for Harassing Chaplain!"* News involving deranged Green Berets were a hot item, guaranteed to boost newspaper sales. It would have been a huge embarrassment for the commander of Army Special Operations Command (SOCOM) to have his executive officer show up on the blotter report. The general would probably fire me, but that was the least of my concerns. The larger concern was that any such publicity would be turned against our efforts to have a Latin Mass held in a Fort Bragg chapel. In 1997 there were very few Catholics who even knew that the traditional Latin Mass was an option, and far fewer still who were dedicated to making that option a reality. There was no Latin Mass authorized by any diocese anywhere in the Carolinas, and for some time we had been sending petitions to anyone who would listen. Most people who would hear about a confrontation between a Catholic chaplain and someone inquiring about a Latin Mass would put the blame squarely on "those Latin Mass people." I had seen how the media was used to manipulate public opinion. Thanks to their efforts, the old rite was largely unheard of and it would remain so for many more years until the election of Cardinal Joseph Ratzinger as Pope Benedict XVI and his 2007 motu proprio *Summorum Pontificum.*

I stood awkwardly by my pickup truck staring into space, wondering what to do next. An enlisted soldier saluted as he passed, jarring me back to reality. But the drive back to my office pulled me back into a dreamy reflection on the past sixteen years, which had landed me here in this confrontation on this day.

Things had always lined up to give me a successful career as an Army officer. Initially, while I was a cadet at West Point, and right up to the time of my commissioning in 1981, I had no plans to make the Army my career. I intended to do the minimum time required and move on to the private sector where I hoped to leverage an engineering degree from a premier institute. Upon commissioning I requested assignment with the 7th Infantry Division at Fort Ord, California, because I had my eye on nearby Silicon Valley, the mecca of high-tech innovation at the time. I went through the Infantry Officer Basic Course, followed by Ranger School. The latter was optional, but I just wanted to prove to myself that I could do it. Other guys went to

Ranger School for career reasons. If you wanted a serious career in the Infantry, you needed to have a Ranger Tab. Not me. I had no Army career in mind. My plan was to go to California, and in my off time start looking into opportunities to work as an electrical engineer in Silicon Valley.

Graduation and Commissioning, May 27, 1981.

Something changed along the way. I began to like the Army. It began to click, and in the end, I never visited Silicon Valley and completely forgot the idea of working in industry. I hadn't exactly been the best student anyway, and now I found that I liked the Infantry, the lifestyle of being a soldier, and the excitement of training. I liked the ranges, the patrolling, the road marches, the field exercises, and drinking an enormous quantity of beer with my friends when all was done. I was naturally good at mission planning, organizing a group of men, navigating with map and compass, and tactical thinking. As a southerner, the outdoor life came easily, and as a Catholic I knew how to handle fear of death. On that last point, had I known my faith better I would have spent far more time in the confessional. I knew very little about the currents of change flowing through the Church that would eventually sweep me into the wake, but God reveals these things to us according to His plan. For the moment I knew very little about my Catholic faith, which nevertheless sustained me, and nothing at all of the revolt against the Church that was raging under the pretenses of compliance with Vatican II.

Not long after being commissioned, I met my future wife, Lorri, in December 1981, while home in Hattiesburg, Mississippi waiting to ship out to Fort Ord. I had a "Ranger Buzz," which is a haircut that leaves very little hair on your head. By some minor miracle the wacky hairstyle didn't bother her and we began to develop a relationship. I spent only two years at Fort Ord, and we remained in contact.

Fort Ord is still there, right next to Seaside, California, but the 7th Infantry Division is gone. It was deactivated in 1994 and later reactivated with a different mission in a different place. At the time I was there, in 1982 and 1983, the Army was attempting a new concept in which groups of newly enlisted soldiers would go through Basic Training and Advanced Individual Training together, spend two years at a base in the continental U.S., and then the entire unit would be sent overseas together. The concept was called Cohesion, Operational Readiness, Training (COHORT), and the idea was that it was better for teamwork to deploy complete units overseas for a year-long rotation rather than having continuous, individual replacements as had happened in Vietnam. This idea was taken from the regimental system used in the UK. My COHORT was C Company, 3/17 Infantry, to be transferred to the Republic of Korea as a unit in August of 1983.

Lorri came out to visit me in California during a brief leave before our unit's departure for the Republic of Korea. By a stroke of Divine Providence, a good friend was finishing SEAL training in Coronado (San Diego) and was getting married the next day to Lorri's former college roommate. We drove down to Coronado, and shortly after attending the SEAL graduation and wedding, we began making our own plans to marry after I returned from Korea. Weddings inspire more weddings. We married within about a week of my return and we left together for Fort Benning, Georgia, where I was to complete the Infantry Officers Advanced Course.

Much more could be said about my assignments, the people I worked with, and the state of the military at that time, but the intention here is to show how I ended up in a unique situation that allowed me to peek behind the curtain and see things that were being hidden from so many other people.

In 1984 the U.S. was engaged in a series of proxy wars with the Soviet Union throughout Latin America. Cuban-sponsored guerrilla groups had toppled the government of Nicaragua and were working to bring down the governments of neighboring countries. There were four political movements in Colombia: the Ejército Popular de Liberación (EPL) and Ejército de Liberación Nacional (ELN) which had both been around since the 1960s; the Fuerzas Armadas Revolucionarias de Colombia (FARC), a Marxist-Leninist group that also had its origins in the 1960s; and the Movimiento 19 de Abril (M-19), a revolutionary nationalist movement that wanted Colombia to open up to more democratic elections. By the end of the decade the M-19 would lay down their arms and have their leader run for President.

Meanwhile a guerilla war was raging in El Salvador where the Farabundo Martí National Liberation Front (FMLN) was striving to overthrow the government and establish a second Soviet proxy beachhead in Central America, along with Nicaragua.

The social inequalities creating instability in Latin America cannot be denied. The abuses of various corporate entities that led to these conditions are well documented. However, to seek to overthrow the governments of the Latin American countries and replace them with local proxies of the Soviet Union was not the solution. The Catholic Church had long condemned both Communism and unfettered capitalism, but Communism was not well understood in the U.S. There were many people, even Catholics, openly advocating for the advance of this noxious ideology. It had been condemned by Pope Pius IX in 1846[1] and in

---

[1] Encyclical *Qui Pluribus*, November 9, 1846.

1864,[2] by Pope Leo XIII in 1891,[3] and by Pius XI in 1937.[4] In 1949 under the pontificate of Pius XII, the Holy Office declared that those who adhere to this ideology are apostates and therefore excommunicated.[5]

I knew very little of these specifics, only that it was wrong, and it was shocking to see how Catholics around me openly embraced Communism. In retrospect this was part of the wider moral confusion that prevailed during the post-Vatican II era. Relativism seeped into the Church and upended everything. All rules were out, there was no right or wrong, and everything was subject to change. I remained one of those millions of people who continued to believe that what was wrong in the eyes of God *yesterday* remains so for the present and into the future. As for the Catholics who were blind to the dangers presented by Communism, I believe they would have seen things differently if they had been given the opportunity to spend a year in South Korea, where the Catholic Church was alive and well, looking across the border into North Korea where the Catholic Church had ceased to exist.

The Demilitarized Zone in Korea was part of a larger picture. After the end of the World War II, a standoff ensued between the "Free World," consisting of the U.S. and Western European nations, and the Soviet Union and their proxies. This Cold War featured low-key conflicts between the proxies of the Soviet Union and the proxies of the U.S. and our allies. Proxies of the Soviet Union always saw the Church as something hostile and dangerous, and by now we were learning about the persecution of the Church in Nicaragua under the Communist Sandinista regime, and the silencing of Cardinal Obando y Bravo. The reality of the Sandinista era was well documented.[6] Under these circumstances the use of the U.S. military to prevent the continued advancement of the communist ideologues seemed appropriate.

---

[2] *Syllabus of Errors*, released with the Encyclical *Quanta Cura* on December 8, 1864.
[3] Encyclical *Rerum Novarum*, on rights and duties of capital and labor, a key element of modern Catholic social teaching.
[4] Encyclical *Divini Redemptoris*, March 19, 1937. For a synthesis of the teaching of the documents mentioned in this and preceding notes, see Peter Kwasniewski, *His Reign Shall Have No End: Catholic Social Teaching for the Lionhearted* (XIII Books, 2025).
[5] *Acta Apostolicae Sedis*, July 1, 1949, under the heading *Decretum*, presented in the form of a *dubium*.
[6] See Martin Kriele, *Nicaragua: Corazón Herida de las Americas* (Konrad-Adenauer-Stiftung, 1986).

Decades earlier, in October 1961, President Kennedy had visited Fort Bragg, North Carolina and had found the U.S. Army Special Forces to be the kind of low-key presence required for the unconventional warfare that would be needed in the Cold War. He gave the Special Forces the right to wear the distinctive green beret, resulting in the nickname that they carry to this day. Concern over what kind of world future generations of Americans would grow up in was enough to motivate me. After the return from Korea, the wedding, and a six-month stay at Fort Benning for the Infantry Officers Advanced Course, I requested assignment to Special Forces. In Spring 1985 I reported to Fort Bragg to go through the six-month qualification course. Upon completion I was assigned to the 7th Special Forces Group and given command of an operational detachment.

I knew immediately that I had found my home in the Army. Our A-Team made several long deployments to Central America, and if I could have remained a team leader for the rest of my career I probably would have. However, after a couple of years of being nearly constantly deployed, I was transferred to 7th Group Staff as the Assistant Operations Officer. Lorri and I still had no children. She used her time during my long deployments to pursue a degree in Physical Therapy at Duke University and to become licensed as a Physical Therapist so that she could have a career and earn a second income. We were still modern Catholics, and we thought that this was what we needed instead of children.

In 1988 I was offered a position in Colombia as an Exchange Officer with the Lancero School, in Tolemaida, about 60 miles south of Bogota. This training center is the Colombian version of the U.S. Army Ranger School, founded with the assistance of U.S. Army Rangers in 1956. Capt. Ralph Puckett, Jr., a Medal of Honor recipient from the Korean War, was the mastermind behind its inception.[7] He was assisted by 1st Lt. John Galvin, who would eventually retire as a four-star general in 1992 after having commanded both Southern Command (1985 to 1987) and European Command (1987 to 1992). He was in Europe wearing four stars when I received the offer to work at the Colombian training center he had founded decades earlier.

---

[7] Charles H. Briscoe, "Colombian Lancero School Roots: Two U.S. Army Ranger officers develop a training program for the Colombian Army," *Veritas*, vol. 2, no. 4 (2006): 30–37, https://arsof-history.org/articles/v2n4_lancero_page_1.html.

SUPREME ALLIED COMMANDER EUROPE
SHAPE, BELGIUM

April 25, 1990

Dear Captain Sonnier,

Congratulations on your excellent article about the Lancero School recently published in *Infantry* . As one of your predecessors several times removed, I think you hit the mark in your description of the rigorous schedule and the pressing real world exigencies which drive such a regimen.

You have no doubt found that your experiences in Colombia and the friendships you are cultivating have enriched your life immeasurably. I can assure you that you will have garnered countless valuable lessons about yourself and your profession, as well as many treasured memories to last a lifetime.

I encourage you to continue studying and writing about your profession, and to keep your Spanish skills honed. The Army and the nation are always in need of articulate spokesmen with broad perspective and depth of knowledge gained through international experience. Your presence at the Lancero School is certainly building the foundation for a successful career as a "Soldier Statesman."

Best wishes,

Armas a discreción, paso de vencedores. ¡Lanceros adelante!

JOHN R. GALVIN
General, U.S. Army
Supreme Allied Commander
Europe

Captain David L. Sonnier
Lancero School
c/o USDAO
Bogotá, Colombia
APO Miami 34038

Tolemaida, Colombia, December, 1988.

This position required that I first complete the Lancero training, which was like going through Ranger School again, only in Spanish and a few years older. I finished weighing probably thirty pounds less than when I started, having survived both malaria and dysentery. I took all due precautions, but the prophylaxis in use for malaria at the time was inadequate for jungle training in the Putumayo region. As for the dysentery, I had been careful to purify my water, but these things can still happen. It was toward the end of the physically demanding first phase. A one-day break was near, so I tried to hold out. I didn't want to miss training to seek the somewhat sketchy frontline medical help. One of my classmates, upon seeing me make repeated trips to the "green latrine," informed me that I had dysentery.

"It's not going away without medication," he said. Then he gave me a packet of Flagyl. "Take it. You can have all of it. I won't need it. I'm leaving tomorrow to go to seminary."

I was dumbfounded. Seminary! I had almost never in my entire life heard of someone going to seminary. I looked at him closely, trying to wrap my mind around it. It was dark and we were sharing guard duty. A guy in *Lancero* training was leaving to become a priest?

"What are you doing here?" I asked, incredulously.

He laughed. "I just found out today that I got accepted. So, I'm leaving first thing in the morning."

Our relief arrived a few minutes later, and I never had the opportunity to speak with him again. Sure enough, the next morning we saw him quietly loading his backpack into a vehicle and leaving.

I completed the training, Lorri arrived in time for graduation, and we were given a place to live in the Colombian officer's housing area. We did the best we could to establish friendships with the people around us. Very few of the Colombians spoke English, so Lorri studied Spanish with a genuine sense of urgency. My work came in training cycles, in which I would be extremely busy for two to three weeks straight, followed by several days off during which we could rest, relax, and travel. Usually, we traveled to Bogota and stayed with friends there. Our mail came to the office of the U.S. Military Advisory Group, which was in the Colombian Army Headquarters in Bogota. After picking up our mail and spending a day going through it, we would make telephone calls to the U.S. It was convenient to make phone calls while we were in Bogota. The base in Tolemaida was quite remote. The Colombian training battalion commander had a microwave phone in his office which he could use to call headquarters in Bogota and I was able to use it from time to time to call the MIL GROUP, or they could use it to reach me. If, however, we wanted to call a commercial number in the U.S. we had to go to the nearby town of Melgar to a "telephone office" building that looked somewhat like a barber shop. We'd give the number to the operator at the front desk, then take a seat. In a few minutes they'd let us know the call was ready and direct us to a booth. In the booth, we'd pick up the handset and begin talking to the party back in the U.S. That was the state of technology at the time.

Over the next year my work at the Lancero School involved marksmanship instruction, patrolling, riverboat operations, rope bridge construction, and related training every day from 4:00 in the morning until late at night for a couple of weeks at a time. An intense cycle of this kind of training was followed by a trip to Bogota. Eventually my wife and I began to venture further as we got to know this magnificent Andean country.

We could not venture far. There were huge swaths of the country where travel was not allowed. We could not go anywhere near Medellin, for example, where drug kingpin Pablo Escobar was already making a name for himself.

One of the Colombian officers told me about the bishop of Pereira, Dario Castrillón-Hoyos. He was admired among the officers for his legendary courage and the excellent example he set as a bishop. He once disguised himself as the milkman to get access to the residence of Pablo Escobar and speak with the drug kingpin and, according to this story, he persuaded Escobar to confess his sins. One rarely encounters this type of courage among prelates. Dario Cardinal Castrillón-Hoyos would eventually end up in the Vatican working with groups of Catholics dedicated to traditional liturgy and practices.

Colombia was a deeply Catholic country; in fact, Catholicism was the state religion. On Sundays when I was home, we would attend Mass in the early evening by walking to a nearby open-air court set up with an outdoor altar and chairs for officers and families. About 200 soldiers stood in formation around us. Attendance was obligatory for them. During Semana Santa—Holy

Week—everything came to a standstill as events from two thousand years ago were dramatically commemorated. We attended a "Living Stations of the Cross" in which soldiers acted out the various events along the way of the Passion and death of Our Lord. Throughout the year there were numerous other reminders that we were living in a Catholic civilization. Once a year all the units on the installation would arrange their weapons on the soccer field for the "blessing of the arms." A priest would slowly walk through the rows of stacked rifles, mortars, and artillery pieces sprinkling holy water on them and giving a benediction for their correct use, invoking divine protection, and reinforcing the soldiers' sense of purpose in defending their country.

On one occasion a young officer I worked with invited us to his wedding. When I asked him whether the wedding would be held in a Catholic Church, he was astonished by my ignorance. "Of course!" he said. "In this country, if an officer in the army wants to marry it must be in the Catholic Church. You cannot be married in any church other than the Catholic Church if you're in the Colombian Army." This was revised by their 1991 Constitution, which ended the period in which the Catholic Church held the status of state religion.

Colombia suffered immensely during this period. There was an outbreak of violence before the upcoming 1990 elections, and three of the leading presidential candidates were assassinated. Carlos Pizarro Leongómez, the commander of M-19 who had demobilized the guerrilla army and transformed it into a political organization, was one of them. On September 2, 1989, one of the two major newspapers in the country, El Espectador, was bombed, with the Medellin Cartel claiming responsibility. It was shortly after this incident that the spouses of all U.S. Military and State Department personnel were evacuated. Lorri spent the last few months of 1989 living with her family back home.

It was during this time that it became apparent that the Cold War was over. We did not have television, and live news was difficult to obtain, but during a brief trip to Panama in 1989 I had access for a couple of days. Watching CNN in a cafeteria at Fort Clayton, I saw young Germans climbing on the Berlin Wall, drinking wine, and breaking off chunks of it for souvenirs. Clearly the Cold War was over. This was all completely unexpected.

The 1989 collapse of the Berlin Wall may not have been a

major emotional event in the lives of some people, but for me it marked the end of a way of thinking and the end of the Cold War. It was pleasantly shocking. I had been completely engaged in the Cold War effort for years, and we lived with a certainty that it would go on for decades. We believed that if we stood our ground, our grandchildren wouldn't have the threat of Soviet world domination to worry about. Now, suddenly, it was over. As I watched the CNN news coverage live from the Berlin Wall, I knew things would never be the same.

Why would we need Special Forces soldiers if the Cold War was over? My world was completely rocked, and now I had to consider other career options.

Back at the *Lancero* School, a few weeks later I received a message to contact the U.S. Military Advisory Group in Bogota. I called the Army Desk Officer, Major Yul Campos, on the microwave telephone. The Army, in its infinite wisdom, made the decision for me. I was going to graduate school to learn about computers and then go on to work as an "automation officer." This obsolete term is no longer used, but an automation officer was a technology expert who worked with the Army's massive information infrastructure.

Technology was playing an increasing role in the world, and the Army sought to increase the number of people educated in Computer Science. I was to find a graduate school and enroll as soon as possible. I flew back home, landed in Atlanta where Lorri picked me up, and found Georgia Institute of Technology on the map. We drove to the campus and found a building labeled "College of Computing" in large bold letters. I walked in, found the dean of the college in his office, and after a fifteen-minute conversation I was on my way to the next phase of life.

I returned to Colombia in early 1990 to wrap up my tour with the Escuela de Lanceros followed by a short stop at the Army Computer Science School at Fort Gordon, Georgia, where I completed a basic course for Automation Officers. In the Fall quarter of 1990, I began graduate studies at Georgia Tech.

Putting all thoughts of the Cold War behind me, I left the world of Special Forces, the Latinos, the great outdoors, and the anti-communist struggle that I had been living in for years and found myself in the world of emerging computer technology. It was mysterious, new, and fascinating. Everything I had learned as an Electrical Engineer was still relevant, but programming had moved to a new level. In the introductory programming class at West Point we had spent an entire semester learning basic constructs such as loops and branches. We would write out our code by hand for the professor to review before typing it into a source code file and uploading it to a mainframe computer to be compiled. That was old school. Basic constructs were now covered within the first few weeks of an introductory programming class. I had never taken a class in data structures, algorithms, or databases, all of which were required for more advanced studies. I had a lot of catching up to do, and many of my classmates had already spent time working in industry. Soon we were on to Analysis of Algorithms, Artificial Intelligence, Operating Systems, Parallel Computing, Programming Language Design, Automata Theory, and Robotics. I learned about Turing Machines only a few years before Alan Turing's contributions to the effort in World War II were formally acknowledged and details were made public. I was finishing graduate school when the World Wide Web project was launching. We regularly used email as students, but to do so you had to know UNIX commands. With the launching of the web, it was not long before email was available to everyone. One of my sisters, trying to figure out what was going on, called one day and asked me, "What the devil is email?" A whole world of possibilities was opening with seemingly infinite opportunities to share information. Studying computer science was an endless process, in which each new discovery led to an understanding of how much there was yet to learn and discover.

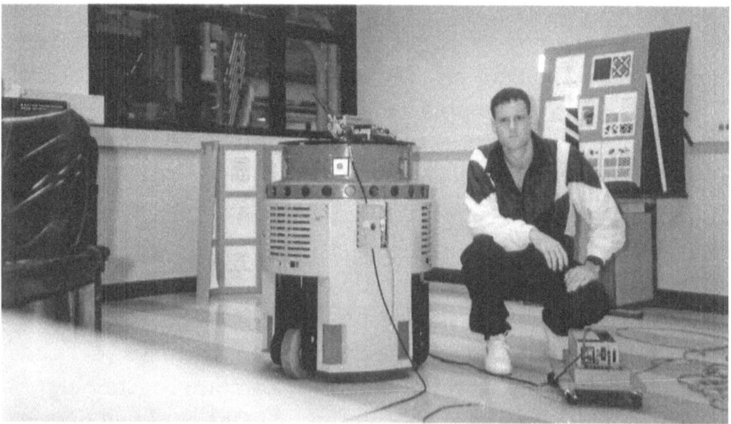

Working on Robotics research at Georgia Institute of Technology under the supervision of Dr. Ron Arkin, February 1992.

Our first child was born while we were in Atlanta. John David gave us a lesson in what our new life would be like, as we both began learning to balance family and work.

I graduated in 1992 with an MS in Computer Science, and we moved to Dayton, Ohio where I was assigned to the faculty at the Air Force Institute of Technology (AFIT), but the program to which I was assigned would soon be dissolved. I was one of four Army faculty teaching classes at AFIT by agreement between the Army and the Air Force. In return the Army would get slots to send young officers to study with us for six months to certify them as "Automation Officers" for their secondary specialty. In retrospect this program made little sense, because if it were possible for an officer to complete two quarters at AFIT, they certainly could finish a graduate degree in two or three more quarters and have an MS to show for it. After I had been there for only two years the program was discontinued and the four of us were reassigned.

This two-year period in Dayton, however, was to become a major turning point in my Catholic life. I continued to completely immerse myself in Computer Science, did research, and even helped publish a paper and presented it at a conference. I taught classes in Computer Networks, Advanced Digital Hardware, and Introduction to Digital Logic. I spent every spare moment thinking and reading about this fascinating new world of technology. The last thing on my mind was the state of our fallen world, events taking place in the Church, and the relationship between these two things.

CHAPTER 2

A RUDE AWAKENING

ONE SUNDAY MORNING IN MAY 1993, AS I SAT in Church daydreaming about some algorithm or an upcoming lab project or some other aspect of my new-found passion, Computer Science, I was shocked into reality by my wife slapping a missalette loudly with her hand. She stuffed it into the book rack, crossed her arms, hissed and glared. I glanced up to see her scowling at the short-skirted lady who was doing the reading. Obviously, I had missed something, which didn't surprise me because I was never able to concentrate on anything going on during Mass.

"What just happened?" I whispered. The lady was still reading.

Without saying a word, Lorri picked up the missalette again.

"Oh, this is ridiculous! Idiots!" she hissed. Then she pointed to the text that was being read. At first, I noticed nothing, but as the lady with the long bare legs read on, I saw that she was altering the gender pronouns of the text as she read. All mascu-line pronouns were changed to feminine. Even when references were to God Himself, the pronoun "He" was changed to "she." This woman had taken it upon herself to change Sacred Scripture. She had been given a role as lector in the liturgy, and she was using her position to alter Scripture to her own liking.

Until this point, working at AFIT had brought me a unique and exciting combination of military life and academia. I woke up every morning looking forward to the day. Our home life was ideal. We lived in a rented house in an old neighborhood in Day-ton, just a short walk from a neighborhood Catholic church. The idea of walking to Mass on Sundays as we had done in Colombia seemed quaint, but we were mildly disappointed with this parish from the first day. My wife and I were so naïve that it had never occurred to us that people could be using the Church to advance ideological or political agendas. We had always believed that those who were in regular attendance at Mass, and certainly those who devoted their lives to the service of the Church as priests, were all there for the same reason: to save their own souls and to help those around them do the same. The problem for me was that I believed that all we had to do was show up on Sundays and Holy Days. Very little thought was ever put into it. Confession was for people who committed egregious mortal sins like murder. I had heard repeatedly that "since Vatican II we don't have to follow a list of rules." That seemed ridiculous. What about the ten commandments? Isn't that a list of rules? "Yes, but the Church is in a new era and we don't focus on what we *cannot* do, but on what we *can* do." No rules anymore. Well, actually there is one rule: you have to show up for Mass on Sundays and of course drop a bit of money into the collection basket. The concept of a "tithe," giving 10% of your income to the Church, was completely unknown to me. I had never even heard it mentioned, so the logic I followed was somewhat like this: if you pay $6 for a two-hour movie, then it would be fair to donate $3 for a one-hour Mass. Whatever the case, showing up for Mass on Sunday was enough; I didn't really pay attention to what was going on around me.

Feminism was the latest rage of the ideological left in the U.S. Women were angry that the priesthood was all men, that deacons were all men, and that on occasion their "right to abortion" was restricted in some small and meaningless way. Some of them were now spelling the word "woman" as "womyn" to avoid using the string of characters that spelled out "man."

We had recently become accustomed to seeing and hearing the angry media attention of feminists. Having grown up in Mississippi, my wife and I were used to hearing critiques of the Catholic Church from the Protestants around us, but it was usually based on misunderstandings of history, theology, or both. As Catholics, we were a minority, so we always felt a bond with other Catholics. But something new was happening. Catholics who continued to participate in the life of the Church were openly expressing refusal to consent to Church teaching. Here we were witnessing a Catholic woman changing Scripture in front of the entire congregation, while the priest sat nearby, apparently oblivious.

There were other problems. This little parish seemed to be a hotbed of dissent. The previous priest had died of AIDS, and the pro-life group in the parish was banned from distributing literature. Many members of the parish were elated over the recent election of Bill Clinton, who had run on a pro-abortion platform and was now forcing the military to accept open homosexuality in the ranks. Joycelyn Elders, the Surgeon General, was hurling accusations at the "male-dominated" Catholic Church, and yet these people still supported the Clinton Administration. Instead of altar boys there were girls; since when had this been allowed?

We had found ourselves frequently changing parishes since our return from Colombia. In retrospect, spending time in a deeply Catholic civilization must have had an impact on us. While I was in graduate school in Atlanta, we had not been satisfied with any of several parishes where we attended Mass. Something always seemed discordant, in some way that I could not quite put my finger on. The discontent was more on my wife's part than mine, as I could eventually find myself daydreaming equally well in any parish she settled on. My wife had been aware of what was going on in the Church for some time, but she did not know what to do about it. She had correctly identified it as a political agenda that was at odds with Catholicism. Unable to find a Catholic solution to the problem, she located a Bible study to attend in downtown Dayton with other women, mostly former Catholics who were now Protestants. The Bible study was based on a fundamentalist understanding of the Scriptures, and the group held a decidedly conservative worldview. They read their English translations of the Scriptures to find meaning in their lives, but when their understanding of the meaning of what they were reading could lead to some Catholic doctrine they would pivot to a less funda-mentalist position and say that at *this* point the Scriptures were meant to be symbolic and not to be interpreted literally.

Despite this intellectual sidestepping, I believe that they were serious and sincere women seeking salvation for themselves and their families. The problem for us was that countering an ideological agenda in the Church with an opposing ideological agenda was not truly seeking a Catholic solution. Lorri began expressing a desire to attend church services with some of these women. This is certainly the path that many well-meaning Cath-olics took in recent decades. They find that their parishes have become a hotbed of leftist activism, and are drawn outside the Catholic Church to find a "conservative" Bible-based Christianity. Eventually they become Protestants. This was not something I was willing to do. She wanted to attend their church to be reassured that Christ didn't lie to us: abortion is wrong because it's murder, the Resurrection really happened, Hell exists and is to be avoided, and marriage is sacred. These are very basic doctrines, but within our parish such basic doctrines were never mentioned, or they were outright rejected. Lorri's idea was that we could attend a non-Catholic church to be reassured about genuine Christian doctrine, then go to Mass to receive the sacraments and meet our Sunday obligation. It was a ridiculous idea, but it illustrates the level of her desperation.

The feminists now had my attention, and I began to engage. Even as disconnected as I was, I knew that changing the Scriptures in the way that I had just witnessed was absolutely off limits. I began to give serious thought to what was going on in the Church and recognized that I was at fault for having paid so little attention. I resolved to do better. I didn't know it at the time, but my wife had been privately praying that I would take the spiritual lead in our family.

The following Saturday a seminar was to be held at our parish. It was announced in the bulletin as a "Life in the Spirit" seminar. Lorri suggested I attend the seminar, and I agreed to do so even though it was going to be an all-day event. All I knew was that it would have something to do with Pentecost, which was two weeks away, and the Holy Spirit.

I arrived at the church, waited, and looked around, making small talk as others arrived. About thirty people were there. The speaker came in, introduced himself, and gave a talk about speaking in tongues. This was something entirely new to me, and from the start I was uncomfortable with it. Speaking in tongues—"glossolalia"—is the practice of uttering "words" that are meaningless. It sounds like random syllables thrown together quickly, as if the speaker is fluent in some foreign language. Those who engage in this practice believe that the speaker is speaking in a language unknown to everyone else. The person who can do this has the "gift of tongues."

I was hoping for an inspiring seminar that would connect me to my Catholic faith. A recounting of the life of some saint would have been helpful. The history of some difficult era in church history, or some straightforward teaching of Catholic doctrine, would have been awesome. Anything genuinely Catholic might have helped me deal with this situation in which my wife was wanting to attend a Protestant church because she wasn't happy about what was going on here. Instead, this gentleman was babbling so fluently that it sounded like a well-rehearsed speech. He claimed he was doing what the Apostles had done when they had spoken to the multitudes in Jerusalem.

At the encouragement of the speaker, others joined in the babbling. The obvious problem that I saw was that no one in the room understood the babble. The claim was that "prayer" spoken through the gift of tongues was in fact being directed to God, and God could understand it, but as I looked around the room, I couldn't see anything but people babbling.

Pentecost Sunday was only a week away. When the Holy Spirit descended upon the Apostles, they were empowered to speak in multiple languages and share the message of the Gospel throughout the world. The speaker emphasized this point repeatedly during the seminar. To make this moment in Christian history meaningful at my parish, the priest would read the Gospel on Pentecost Sunday in English, while others would simultaneously read it in several other languages. This type of experiment with the liturgy was common in 1993. Very few people thought it disrespectful to play such games with the prayers and rubrics of the Mass.

At some point a priest joined the seminar, led a brief discussion, and began almost immediately speaking in tongues. Whether he spoke to God in a language that only God could understand or he just babbled, I don't know. After a short break we spent some time in "prayer," with everyone else in the room doing the babble thing while I silently prayed the only prayers I knew: the Our Father and the Hail Mary. We then read some passages from the Bible and broke into small groups. Within each small group there was a leader assigned to encourage all of us, one at a time, to "speak in tongues." Everyone in the room did this except for me. An elderly woman who had been the only other holdout finally gave in and, to cheers and clapping, she babbled. For some reason the group pressure left me ashamed of myself, but I just couldn't do it. It would have been dishonest. I could babble random syllables to please the crowd, but I knew that it would mean nothing, and offering a false prayer to God for the sake of peer pressure seemed like it would be asking for trouble. We had a brief reception, during which people avoided me, so I snuck out the door and walked back home in a daze, embarrassed, but quite certain that I had done the right thing.

CHAPTER 3

A SPIRITUAL AWAKENING

THE LIFE IN THE SPIRIT SEMINAR HAD LEFT ME with more questions than answers, but I put the confusion aside and tried to focus on my teaching and research at AFIT. By now I had connected with other faculty members, and a small group of us were meeting weekly for lunch. We met at a different restaurant or cafeteria every time, and we would usually not discuss work. A few days after the seminar we had one such meeting, and only two of us showed up. As a result, I was having lunch alone with an Air Force officer named Will, who I knew to be an exemplary Catholic.

This was a rare opportunity. Will was a bit older than I was, a major while I was still a captain, and I looked up to him for his knowledge and insight. Once, he had been a fallen-away Catholic, but when he returned to practicing his faith he acquired a deep knowledge and understanding of all things Catholic. It was always enlightening to hear him speak of Church history or some point of doctrine.

Will set an excellent example by the way he lived his life. During Lent he refrained from eating altogether, but nevertheless joined us for the company. He would often share some pearls of wisdom from St. Thomas Aquinas or one of the Church Fathers. For example, at one point our lunch discussion turned to some question on when it is morally acceptable to engage in armed conflict. I had studied the subject at West Point and later at the Infantry School at Fort Benning, but it was always presented as deceptively complex. Will explained the precise and measurable terms in which St. Thomas had laid out the criteria for "just war."

Since we were alone that day I shared my seminar experience, hoping he would be able to give me some help with my confusion. I told him about the "speaking in tongues," and that my priest was planning to have the Gospel read simultaneously in four different languages the following Sunday. He stared at me for long enough to make me uncomfortable, then gave a rather exasperated sigh.

"The problem with 'speaking in tongues' is that there is no interpretation," he said, almost bored.

"I know. It sounds like random meaningless syllables."

"I'm telling you for the last time. You should go to the Latin Mass."

Yes, he had suggested this several times already, and for one reason or another I had never really heard him. Maybe I had not taken it seriously because we already had a parish, and we had deliberately selected a house to rent that was close to it. It was conveniently located, so I didn't want to think about going elsewhere.

But I was all ears now. A Latin Mass! His suggestion evoked images from my youth: a priest facing a high altar, clouds of incense hovering over him, stained glass windows, bells, music, silence, and prayers whispered at the altar.

Lorri was leaving the next day with the baby to spend a week or two with her parents on the Mississippi coast. She would not even be home on Pentecost Sunday, so this was a convenient time to take Will up on his suggestion. I wouldn't have to worry about running late, the baby crying, or having to look for a good place to park. I could do a reconnaissance by myself, and maybe it would turn out to be something good for us. We were ready for anything that was both Catholic and an improvement on our current situation.

I thanked Will for the kind invitation and wrote down the directions. He suggested that I arrive fifteen minutes early for the Rosary. This was a devotion that my family had sometimes practiced when I was young, and that I had not seen practiced in a church for many years. I already felt drawn to this parish.

Holy Family Catholic Church in downtown Dayton looked like any typical Catholic Church built before the 1960s and 1970s. In other words, it looked like a church. As I entered, it also felt like the Catholic Church I vaguely remembered from my childhood. There were no felt banners, there were no lay people running around at the altar preparing for their various ministries, and there was no noise aside from an occasional cough. The only babbling came from a baby in a pew somewhere in front of me. The women were all modestly dressed and veiled. Someone began leading the Rosary, and I heard a few familiar thumps as kneelers were dropped.

Perhaps the most significant date of my life was Pentecost Sunday, 1993. I was unprepared for what followed. The Rosary having concluded, the priest processed in wearing vestments of a type that I hadn't seen in ages, following altar boys wearing what I now know to be cassocks and surplices. It was when the choir began singing the *Vidi Aquam* that I was instantly transfixed. I had not heard Gregorian chant since I was a young child, but it was nostalgic and familiar. It sounded magnificent with the subtle organ accompaniment, and it brought back memories of how the Church was when I was in the first grade. The nuns who taught at my school sang at High Mass, and it sounded like this. As the priest began the Prayers at the Foot of the Altar, the Choir began singing the *Kyrie* of the *Mass of the Angels*, or Mass VIII. I quickly found this sung prayer in a small red missalette in the pew, with the text of the Mass on the left page and the translation on the right.

Kyrie eleison, Kyrie eleison, Kyrie eleison.
Christe eleison, Christe eleison, Christe eleison.
Kyrie eleison, Kyrie eleison, Kyrie eleison.

As the *Gloria in excelsis Deo* began I got chills. There was something supernatural happening. I didn't know any Latin at all, but that minor detail seemed to be of no consequence. The prayers were directed to God, and the translations were readily available. Unlike the indecipherable babbling noises that I had heard the previous weekend which were supposedly directed to God, it was now clear that these prayers really were directed to God, weighted with centuries and millennia of meaning. The priest and the congregation all faced in the same direction as he led the congregation in prayers. To God.

The priest chanted the Epistle and Gospel. None of it was delegated to the laity. There were no short-skirted women flitting around the altar. There were no announcements of birthdays, no suggestions to turn and meet our neighbors in an awkward "sign of peace," and no requests for all visitors to stand. There was no applause. There was no guitar music.

Within moments I knew I had found my spiritual home. This

was absolutely, without a shadow of a doubt, exactly where I belonged and God, in His infinite mercy, had led me here.

The priest gave a homily that didn't include the slightest trace of ideology. In the 1990s it was typical for priests to use the pulpit to condemn the sins of bigotry, intolerance, and not being open to change. It was acceptable for priests to elaborate endlessly on the same political narratives that were already being trumpeted by the media, and hearing them repeated from the pulpit always put me to sleep. We all knew by then that abortion was murder, but we had never heard a priest condemn it in the setting of a homily. I breathed a sigh of relief because today I wasn't going to have to tune out some secular political agenda.

I remained transfixed until the end of Mass. When it ended, there was no rush for the door, there was no talking, and it was several minutes before people began quietly leaving. I waited outside the church for Will and his family. As it turned out, they were all members of the choir, and I met one or two other choir members that day. Will invited me to lunch, and since my wife was out of town and I was full of questions, I gladly accepted the invitation.

Will and his family lived in officer housing at Wright Patterson Air Force Base. He fired up the grill on the back porch and began cooking while I peppered him with questions. My curiosity was overwhelming, and he probably felt that I was interrogating him. Each answer he gave led to a dozen new questions.

"Didn't some French bishop got excommunicated for saying Mass in Latin?" I asked.

"You're probably referring to Archbishop Lefebvre. No, he didn't get excommunicated for offering Mass in Latin. Any priest or bishop can do that. The pope often says his Daily Mass in Latin. The French archbishop was excommunicated because he consecrated several bishops without having permission from the pope to do so."

"Didn't Vatican II do away with Latin?"

"No. The Vatican II document on the liturgy clearly states that Latin and Gregorian chant should remain in the Mass."

"Really? So, why do we never see that?"

"Within the Church there are Modernists. They basically hate the idea of people praying in Latin, so they removed it by convincing everyone that it's forbidden."

"What is a Modernist?"

"If you ask me, they're heretics. There was a heresy defined as 'Modernism'…"

"Does the Church still teach that there is such a thing as heresy?"

"Yes, of course."

"So, why do we never hear sermons condemning it?"

"Good question," he replied. "Why do you never hear sermons condemning abortion?"

That was a good question, and it related directly to the unsettled state of mind that my wife was in. Some of the most serious moral evils of the day were never publicly condemned from the pulpit. Any such list would have to include abortion. Yet it was never mentioned. Why were the pro-life people at my parish treated so badly? They were genuine, dedicated, sincere activists who believed in the truth of their cause. I couldn't get as worked up and emotional about it as they were, but there was no doubt that they were on the right side of the issue. Nevertheless, discussion of the issue was not allowed, nor were they allowed to distribute literature.

Will made a generalization: "Usually you get good Catholic doctrine and good explanations of what the Church teaches if you attend the traditional Latin Mass. Otherwise, it seems to be hit-or-miss." Over the years I've found this generalization to be helpful and accurate. My mind was made up. We would be attending Holy Family from now until my assignment at AFIT ended.

Over the course of the next couple of hours I learned several things. Although the Mass had been codified after the Council of Trent (1545–1563) into a form very much like the one we had attended that morning, minor changes had been made over the centuries. People sometimes referred to the Latin Mass as the "Tridentine" Mass, meaning the Mass of the Council of Trent, but that was not entirely accurate. The term "traditional Latin Mass" was a better descriptor, since it embraces a much longer history that stretches back over a thousand years before the Council of Trent. There had been minor changes right up until 1962, but after that edition of the missal, liturgical changes had become quite radical. The "New Mass" that had emerged after the Second Vatican Council, which took place from 1962 to 1965, was more appropriately called the *"Novus Ordo."* This was what was currently found throughout the Western Church, except where the traditional Latin Mass was available by "indult," a formal term for permission given by the bishop.

In 1988 Pope John Paul II asked all the bishops to allow the old Mass when it was requested. Some of the bishops complied, but usually they simply refused permission. As Will explained it, there was a Latin Mass here in Dayton because the bishop permitted it in accordance with the guidance from Pope John Paul II.

As he continued, I found it difficult to believe that everything he was telling me was true. The Second Vatican Council had never required that Latin be removed from the Mass, and Gregorian chant was also to have been retained. Really? He sent one of his daughters to fetch a large, yellow book with "The Second Vatican Council" written on the cover, and read directly from it:

The use of Latin is to be preserved in the Latin rites. (Constitution on the Sacred Liturgy, no. 36)

The treasury of sacred music is to be preserved and fostered with great care. (Constitution on the Sacred Liturgy, no. 114)

The Church recognizes Gregorian chant as being specially suited to the Roman liturgy. Therefore, other things being equal, it should be given pride of place in liturgical services. (Constitution on the Sacred Liturgy, no. 116)

Latin and Gregorian chant were both supposed to have been preserved? Until that day, I had not heard the first word of Latin or the first note of Gregorian chant in any church that I regularly attended since I was a child! Instead, we had been told repeatedly that Vatican II changed the Mass from Latin to the vernacular, that is, the local, commonly spoken language. We had been told this repeatedly. My head was spinning, and a dreadful thought began to cross my mind. Had we been lied to for all these years? Was it possible that Catholics knew their Faith so poorly that we had all been tricked? Unless someone had access to the actual documents of the Second Vatican Council, how would they know?

Perhaps Will was lying. I watched him as he fiddled absent-mindedly with the grill. It didn't seem possible that he was making this up. He had been so knowledgeable about all things Catholic. But either he was lying or most of the hierarchy of the

Church had been lying to the faithful for years. If what he was saying was true, why was I hearing it from him, and not from a priest or bishop? Or why wasn't it in Catholic magazines and newspapers?

Will explained: "It was the way these documents were written. That was the catch. There was ambiguity built into these declarations leaving loopholes big enough to drive a truck through."

He picked up the large book again. It was about the size of the Dayton phone book.

"Let me give you an example: 'The use of Latin is to be preserved in the Latin rites, but since the use of the vernacular, whether in the Mass, the administration of the sacraments, or in other parts of the liturgy, may frequently be of great advantage to the people, a wider use may be made of it, especially in readings, directives and in some prayers and chants.' See? There's a loophole. The Mass is supposed to be in Latin, but … and that word 'but' gives them license to do whatever they want."

From that moment on, I have never been able to reconcile what was written and proclaimed in the Second Vatican Council document on the liturgy with what actually happened. It made no sense then, and it makes no sense now. For centuries people are praying in Latin every Sunday, and then suddenly the highest authority in the Church says, "You can use more vernacular language in your prayers if you want." And then the bishops take that new guidance and proclaim: "Latin is now forbidden!" It's all a lie. There is no way that anyone could interpret the clear text of the Second Vatican Council document to be a call for the elimination of Latin from the liturgy, but that is what had been done.

Will continued, "Here's another example: 'The Church acknowledges Gregorian chant as specially suited to the Roman liturgy: therefore, other things being equal, it should be given pride of place in liturgical services. But other kinds of sacred music, especially polyphony, are by no means excluded from liturgical celebrations, so long as they accord with the spirit of the liturgical action...' See? Another loophole. Other kinds of music can include guitars or bongo drums."

"Polygamy? What?"

"No, polyphony. It's a type of music, usually in four parts that go in different directions, but with the same text, and the parts harmonize with each other. You've probably heard it in recordings or concerts. But probably not at Mass."

So, according to the highest authority in the Church, Gregorian chant was best for the liturgy, but other types of music could be allowed. That had somehow been interpreted as a call for the complete elimination of Gregorian chant. This wasn't someone taking advantage of a loophole, it was an outright lie.

I was getting a headache. It was too much. How was this even possible? There had to be more to it. Intellectual honesty requires humility, and obviously there were some additional facts bearing on this situation that I would have to research. Now that I was awake, I resolved to do whatever was necessary to find out what was going on.

As for the Catholic magazines and periodicals that had all along been reporting on what was happening as events unfolded after Vatican II, they did exist. Will gave me some copies of *The Wanderer* and *The Remnant* as I was leaving.

I was still skeptical, and I had a headache.

CHAPTER 4

HOW DID WE GET HERE?

THE NEXT SATURDAY MY FAMILY WAS BACK IN Dayton, sunburned from the beach. I was prepared for some pushback from Lorri when I suggested that we attend Mass at Holy Family the next day, but she was quite willing to give it a try.

We arrived early enough for the Rosary and sat close enough to the back that we could get out the door if the baby made a scene. As Mass began, I showed her where we were in the red missalette, but after a few minutes she no longer needed my help. A few minutes later I looked over and saw her wiping tears from her eyes. Once again, I was overwhelmed by the solemnity and beauty of the Mass.

On the way home, she asked, "Why have we never heard that it is still possible to attend a Latin Mass?"

I was now armed with some new information, but it was not enough to answer that simple question. Why indeed? One could find Mass in any language easily enough, so why not Latin? It seemed it should have been more prominently featured among the options for contemporary Catholics. I was also disturbed by the emerging contradictions. We had always been told that after Vatican II the Mass was to be exclusively offered in the vernacular, yet clearly that was not what was intended by those who signed the document that described how the liturgy was to be revised: "The use of Latin is to be preserved in the Latin rites," according to the Constitution on the Sacred Liturgy. The actual outcome had not only been a loss of the use of Latin, but also a degrading of the integrity of the Mass. If the use of the vernacular was the goal of whoever had revised the liturgy, why wasn't the Mass directly translated, and otherwise left alone? Highly significant and beautiful prayers were removed, such as the Prayers at the Foot of the Altar at the beginning, and the reading of the Last Gospel (John 1:1-14) at the end. The entire Offertory was gutted. There were multiple options for

Eucharistic prayers, and instead of a one-year cycle of readings there were different Scripture readings every Sunday in a three-year cycle.

As I continued to learn, I discovered that the Mass which had existed with few changes or differences since well before the Council of Trent had been entirely replaced in 1969. But *had* it been replaced? Was the old Mass now forbidden? Obviously not, because it was available by "indult" in Dayton, Ohio and a few other places. Very few, though. I contacted the "Coalition in Support of *Ecclesia Dei*," an organization in Chicago that had printed the small red missalettes we found in the pews at Holy Family. They mailed me some information that included a list of locations in which one could find a traditional Latin Mass in a diocesan church under the indult of a local bishop. According to this list, the traditional Latin Mass was virtually nonexistent in the Deep South. There were a few exceptions; it was available once a month in the Cathedral of Biloxi, Mississippi, and on Sundays in New Orleans.

The New Mass, named the Novus Ordo, did not come about as the direct result of Vatican II. It was composed by committee a few years later. This committee included Protestants and revolutionary Catholics, inspired and led by Archbishop Annibale Bugnini, an exiled radical under suspicion of Freemasonry.[1] In April 1976 Tito Casini, an Italian Catholic writer, publicly brought these accusations to light. Archbishop Bugnini denied the allegation that he was a Freemason, but Pope Paul VI assigned him to a titular post in Iran within just a few hours of being presented with a dossier on his Masonic affiliations and activities.[2]

---

[1] See Tito Casini, *Nel Furno di Satana* (Carro di San Giovanni, 1976), 150. For more detailed accounts of the postconciliar liturgical reforms, see Yves Chiron, *Annibale Bugnini: Reformer of the Liturgy*, trans. John Pepino (Angelico Press, 2018); Boniface Luykx, *A Wider View of Vatican II: Memories and Analysis of a Council Consultor* (Angelico Press, 2025); Louis Bouyer, *The Memoirs of Louis Bouyer: From Youth and Conversion to Vatican II, the Liturgical Reform, and After*, trans. John Pepino (Angelico Press, 2015), 218-30; Peter Kwasniewski, *Close the Workshop: Why the Old Mass Isn't Broken and the New Mass Can't Be Fixed* (Angelico Press, 2025), 3-26, 47-68, 98-101.

[2] Michael Davies, *Pope John's Council* (Angelus Press, 2007), 266-67; see also the articles mentioned in Kwasniewski, *Close the Workshop*, 51n7, which mention the more recent revelations made by Fr. Charles Murr, secretary to the cardinal (Edouard Gagnon) who had been appointed by Paul VI to investigate Freemasonry in the Vatican.

The discovery was too late. Years before this exposure of his Masonic ties, Archbishop Bugnini had already put the pieces in place to usher in the most radical changes to the liturgy that the Church had ever undergone, and had it approved by Pope Paul VI. Liturgies are not designed by committees in either the Catholic or the Orthodox Church. They develop slowly and organically over centuries. *Sacrosanctum Concilium*, the very document used to justify this radical revision, proclaimed: "There *must be no innovations* unless the good of the Church *genuinely and certainly requires them*; and care must be taken that any new forms adopted should in some way grow organically from forms already existing."[3]

What had once been considered holy and sacred was old and tiresome, in the minds of the "reformers," and had to be replaced at all costs. They managed to push the radical revision through, but instead of ushering in a great new era it was a disaster. Millions of Catholics, including consecrated religious, lost their faith and quit attending Mass altogether. In Quebec, Mass attendance plummeted from more than ninety percent to under ten percent. It was worse than most people could comprehend. Vocations to the priesthood and to religious orders slowed to a standstill. Orders of priests and nuns began dying out.[4] Those who were behind the liturgical revolution doubled down by introducing even more bizarre and radical novelties at a dizzying pace. To hide the damage done, false narratives were advanced through the media, both Catholic and secular. The Church was amid a "springtime"! Vatican II required all these ongoing "reforms," and they were turning out "better than expected." The reality was that the Holy Church was in a state of absolute liturgical and doctrinal chaos. The most egregious innovations found in the Novus Ordo had not been prescribed anywhere in the document *Sacrosanctum Concilium*. There was nothing about abandoning the high altar, or the priest offering Mass facing the congregation.

---

[3] *Sacrosanctum Concilium*, no. 23.

[4] The massive decline in numbers was thoroughly documented by Kenneth C. Jones in his *Index of Leading Catholic Indicators: The Church Since Vatican II* (Oriens Publishing Company, 2003). More recent analysis has been conducted by sociologist Stephen Bullivant. For updated statistics and projections, see the work of James Green at his Substack *Grain of Wheat*, e.g., "Quantifying the Catholic Revolution" of February 22, 2026.

There was no mandate to stop receiving Holy Communion at the altar rail, to receive Communion under both species of bread and wine, or to receive communion from the hands of unconsecrated lay people. There certainly was no mandate to abandon the use of the unifying Latin language and Gregorian chant in the liturgy. We, the laity, had been lied to about everything. And those pushing for the eradication of all Catholic tradition knew most Catholics wouldn't go read the Vatican II documents for themselves. Once the new Mass was promulgated, they would just think that surely the pope must be right when he attributed it to Vatican II. For those few who had access to the original documents and raised questions about the radical innovations, the narrative became that it was done in the "Spirit of Vatican II."

My friend, Will, guided my efforts as I researched. He told me about the mistranslations of the Latin into the vernacular. For example, in the original English edition of the Novus Ordo, the text of the consecration was translated: "This is my blood, which is shed for you and *for all* for the forgiveness of sins..."; but the Latin *"pro vobis et pro multis"* conveys a different meaning, since the word "multis" means "many," not "all." It may seem like an insignificant point to the casual observer, but "for you and for *all*" is not the same as "for you and for *many*."[5] The original Latin text indicates that Christ's redemption saves *many*—but not *all*—since some people reject salvation. Among those who had noticed this error, the concern was that it could lead people to accept the heresy of universal salvation. In fact, by this time, nearly twenty-five years later, many Catholics already believed in universal salvation. It's doubtful, of course, that this mistranslation alone was the single cause of the widespread belief in universal salvation. Some prominent theologians, such as Hans Urs von Balthasar, had suggested that universal salvation was a possibility. I had personally heard the mantra "we no longer believe in lists of rules" many times while I was still quite young, and I had been left with the distinct impression that all you really had to do to save your soul was to attend Mass on Sundays and Holy

---

[5] In 2011, after four decades, the mistranslation was finally corrected in a new edition of the English Novus Ordo; but still to this day in many other languages the same erroneous phrasing remains, in spite of Benedict XVI ordering that it be corrected—including in his own country of Germany.

Days. By 1993 Catholics also widely believed that the concept of Hell was an outdated, medieval superstition. This widespread misunderstanding had led to our churches becoming something like social clubs. The parish we attended up until a couple of weeks before was a social club for liberal Catholics. The reason we did not fit in was because we were not liberals, but fortunately our wandering had led us to a Latin Mass instead of a different social club for conservative Catholics. Traditional and conservative are not the same thing.

Will pointed out other mistranslations: The Nicene Creed, in Latin, begins with *"Credo in unum Deum..."* which translates to "I believe in one God...." At that time, attending any English Novus Ordo Mass one would find that the creed they were reciting began with *"We* believe in one God..." When reciting the Creed, you can only speak for yourself. You cannot speak on behalf of anyone else. This error in translation fit nicely with the current overemphasis on the communal aspects of the Mass and the notion of the parish as a social club. These mistranslations were eventually rectified, but in 1993 any such corrections were still many years away.

Some of the other changes in the Mass had absolutely no basis in *Sacrosanctum Concilium*. There was no mention of reversing the direction in which the priest is facing during Mass, for example. Throughout the history of the Church, it had been understood that the priest faced *ad orientem*, to the East. The building was constructed, whenever possible, such that the high altar of the church was at the East side of the church, and it was there that the priest stood as he led the congregation in prayer. Now the high altars were neglected, or sometimes even destroyed, and a floating altar was placed such that the priest could face the congregation. This posture had no basis in any documents of Vatican II. Nor had there been any mandate to remove altar rails from churches. Nevertheless, the "Spirit of Vatican II" was always invoked to both justify changes and silence dissent.

In 1993, at the time I was finding out the extent to which we had been lied to about Vatican II, female altar servers, or "girl altar boys" as we called them, were strictly forbidden and considered a grave abuse of the liturgy. This practice was forbidden in at least two documents that had come from the highest authority of the Church after Vatican II:

> In conformity with norms traditional in the Church,
> women (single, married, religious), whether in churches,
> homes, convents, schools, or institutions for women, are
> barred from serving the priest at the altar.[6]

> There are, of course, various roles that women can per-
> form in the liturgical assembly: these include reading
> the Word of God and proclaiming the intentions of the
> Prayer of the Faithful. Women are not, however, permit-
> ted to act as altar servers.[7]

These instructions were disregarded by most of the Catholic
Church in the U.S. Now that I knew that it was wrong, I was
quite irritated by this clear act of rebellion. The priest at our
previous parish must have known that this practice was not
allowed, even if his congregation didn't know.

I showed the instruction from *Inaestimabile donum* to my
wife. She chuckled. "A couple of months ago I complained to
you about this, and you didn't seem to think it was any big deal!"

"I didn't know any better then. Anyway, if women are allowed
to read the 'Word of God' that seems like a more important
role, so what difference would it make if another woman is also
serving as an altar boy?"

"What are women doing reading the 'Word of God' anyway?"
she asked. "This is contrary to St. Paul's teaching that women
are to remain silent. What does this mean? Do they read the
Epistle and the Gospel? Just the Epistle? Just the Gospel? What
is going on? Who is in charge?"

"The pope is in charge, John Paul II, and he says they're not
to wear vestments and pretend to be altar boys," I said. "Anyway,
don't worry about it. We're not going to Mass there anymore."

I heard no response, so I looked her way. She was staring at
me in disbelief.

"That's it? We're just going to abandon all our friends there
and start attending Mass at Holy Family? Just like that?"

She had been praying for me to take the lead as the spiritual
head of the family, and I think she was beginning to regret it.

---

[6] Sacred Congregation for Divine Worship, Instruction *Liturgicae instaurationes*, September 5, 1970, no. 7; Latin text here: www.cultodivino.va/content/dam/
cultodivino/documenti/Istruzione-Liturgicae-instaurationes.pdf.
[7] Sacred Congregation for the Sacraments and Divine Worship, *Inaestimabile
donum*, April 17, 1980.

"Yes, that's exactly what we're going to do. Or continue to do. We haven't been there in several weeks."

"But what about our friends there?"

"We will invite them to join us."

Between periods of studying and reading I found time to reflect. How many Catholics had I known who had simply quit living as Catholics? There were so many that I simply couldn't recall them all, but it was certainly the majority. I had spent a summer working in the oil industry in Louisiana. A coworker twenty years older than me had gone to Catholic school in the same town where I grew up. He had been an altar boy.

"Ah quit goin ta Maaaass wheen they took th'Latin owt of it. I couldn't staaand all that backslappin' b—ll s—t," he explained in his deep Mississippi drawl.

So many of my classmates at West Point who were Catholic never attended Mass. One classmate said he wouldn't go to Mass because he was a "Roman Catholic," and his family only went to certain churches in which they didn't do the "handshaking kumbaya stuff," and therefore he wouldn't attend Mass at the Catholic Chapel at West Point. Another classmate, and close friend, left the Church over a particular incident. He was like me in many ways; we were both from Mississippi, and we were both Catholic, but he was far more politically astute and sensitive to agendas. For the first year we used to walk to the Catholic Chapel together on Sundays. Then, sometime early in our second year a couple of female cadets were caught in a compromising situation. They were lesbians, and it took very little time for the story to circulate throughout the entire Corps of Cadets. The two were removed from campus the next day, but the following Sunday our priest admonished us severely during the homily for judging others and he lectured us on the "virtue of tolerance." My friend was deeply hurt to see a priest using his position to advance an agenda that could in no way be reconciled with his core beliefs as a Catholic. He was livid. I sat through the angry lecture in mild amusement, but while we were walking back to the barracks my friend unloaded on me. I had never seen him like this. He never returned to Mass, and he soon became a Protestant.

I remembered a shy girl that sat next to me on the school bus during high school. She didn't like having to go to Mass anymore. "I can't stand the handshake thing! Joe has a crush

on me and he always sits in front of me at Mass so that I have
to shake his hand." She left the Church as soon as she was no
longer living with her parents.

My own father quit attending Mass for a while after the series
of neverending radical changes began. He was from the southern part of Louisiana which was French and deeply Catholic.
Some of what was going on in the Church must have offended
his sensibilities, even though he was quite liberal. His French-
speaking family was patriotic as well as Catholic. At sixteen
years of age he had joined the Marine Corps during World War
II, following five of his brothers who were already serving. Three
brothers served in the European Theatre, one in the Pacific,
and one served in Alaska. By the time he completed boot camp
the war was over. After serving out his time in the Marines he
spent some time discerning whether or not he had a vocation.
Before Vatican II it was typical for young Catholics to discern
whether they had a vocation as a priest or with some religious
order, so he spent a year with the Christian Brothers in New
Orleans. He discerned that he had no vocation and moved on to
college, eventually pursuing a doctorate in Education where he
found his niche with the left-leaning crowd. But what was happening with the Mass by 1970 was too much even for his liberal
sensibilities, so he would stay at home while my mother—the
convert in the family—brought the six of us to Sunday School
and Mass. Eventually my father began attending Mass again,
and a few years later, my family joined a Rosary group. When
I was at West Point, he spent a few days with me in New York
and brought me to Mass at St. Patrick's Cathedral. Part of the
Mass was in Latin, which I had not heard in a long time, and
never did again until I started attending Mass at Holy Family. He
explained to me the importance of Latin as a universal prayer
language. He said that since it was a "dead language," the meaning of each word was fixed, as were the doctrines of the faith
expressed in it. It could not be manipulated, and the meaning
of words could not be changed. The universal prayer language
made the Mass accessible to everyone. People from different linguistic backgrounds throughout the world could pray together
in the universal language of the Church. As I now recall that
Sunday and my father's simple and accurate explanation of the
Catholic tradition of praying in Latin, I began to understand

what he must have gone through when the Mass he had grown up with was snatched away from him.

I thought back on the Catholic school that I attended in first grade, academic year 1965-1966. In a few short years a mass exodus of female religious would decimate Catholic schools in America, but in 1965 in Greely, Colorado each class had a sister that taught all subjects, shepherded us around campus in single-file, and disciplined us when we were unruly. My class had Sr. Dianne, who taught us to read, write, add, subtract, and avoid losing our souls. The sisters wore a full habit, so we had no idea what color her hair was, but she told us that she had been born with purple hair. We knew it was a joke, and everyone laughed. But she also told us that there was a "big council going on in Rome, and after it's over we will be allowed to show our hair. You'll see! It's purple." More laughter. She also warned us that when she got to heaven, she had better see each and every one of us there, OR ELSE! Now I wondered whether she had left her vocation in the chaos that followed the "big council in Rome." We had been able to attend this excellent elementary school despite my family's financial hardships, thanks to the sacrifices these women made in a religious order dedicated to education. After the "big council in Rome," Catholic schools were increasingly forced to hire people who needed to make a living, and they gradually became too expensive for anyone but the relatively wealthy.

There was more going on than the radical revision of the Mass. As a young officer I had already become painfully aware of something that many people are still unaware of to this day—the fact that media news cannot be counted on for accurate reporting. I became most aware of it while working in Central and South America. Operations that I participated in were misrepresented. Events with which I was intimately familiar were presented to the American people with obvious bias for the purpose of building a narrative in support of a political objective. There was still no internet for alternative information to leak through. Everything consumed as "news" came from a limited number of sources. Now I was questioning whether, and to what extent, events involving the Church had also been misrepresented. I felt naïve for not having considered that the same media that was misleading us about everything else might also lie about events taking place in the Church.

Too many people trust what they regard as honest reporting from the media. The early twentieth-century historian, philosopher, and mathematician Oswald Spengler described the process in which people are manipulated by mass media:

> The Press to-day is an army with carefully organized arms and branches, with journalists as officers, and readers as soldiers. But here, as in every army, the soldier obeys blindly, and war-aims and operation-plans change without his knowledge. The reader neither knows, nor is allowed to know, the purposes for which he is used, nor even the role that he is to play.[8]

Obviously, if it's done in the realms of domestic and international politics it can similarly be done to mislead the world of Catholics into a false understanding of Vatican II. Spengler echoes the question asked of Our Lord by Pontius Pilate, and then partially answers it: "What is truth? For the multitude, that which it continually reads and hears."

If the multitude is repeatedly told that "Vatican II changed the Mass from Latin to the vernacular" then that becomes truth. Nobody believes you when you quote directly from no. 36 of *Sacrosanctum Concilium*, where it says "the use of the Latin language is to be preserved in the Latin rites," or no. 54, where it says "steps should be taken so that the faithful may also be able to say or to sing together in Latin those parts of the Ordinary of the Mass which pertain to them." Frequent repetition had definitely steered most Catholics toward the narrative that we were not supposed to be praying anymore in Latin, period. That still didn't explain why the bishops didn't clarify the matter. Most bishops had their own media under their control, for disseminating information within their dioceses. Were they cowed into submission by peer pressure, or were they deliberately propagating false narratives?

I couldn't wait to talk with Will again. Occasionally the two us would meet for lunch, so that I could once again pepper him with questions. I admired his patience with me.

"So," I asked on one of these occasions, "Latin and Gregorian chant were both supposed to continue to be part of the Mass,

---

[8] Oswald Spengler, *Decline of the West*, vol. 2: *Perspectives of World-History*, trans. Charles Francis Atkinson (Alfred A. Knopf, 1928), 462–63.

and this wasn't done. There was no mandate for removing altar rails, reversing the direction of prayer, and basically destroying the beauty of our churches, but this was done anyway. Why didn't Catholics protest?"

"How were they to protest? To whom could they bring their concerns?"

"I don't know ... their priests, or maybe their bishops," I replied.

He didn't answer right away, apparently searching for words that wouldn't scandalize me. "I'm going to give you an uncomfortable fact. Most bishops in the U.S. don't want anything at all to do with the Latin Mass. There are a few exceptions, but for the most part they want to avoid confrontation."

"Why not? It's so beautiful. The prayers are beautiful, and meaningful. The music is beautiful. Why would there be any confrontation?"

His answer was disturbing. "The bishops were led to believe that people wanted a radical makeover. Eventually some bishops may give permission in one location in their diocese for a Latin Mass, in a way that's so restrictive that you can't possibly expect a community to grow out of it—like 6:30 in the morning, or in a drive-by shooting district on a Saturday night, or one Thursday night per month. There are a few bishops who recognize that the post-conciliar reforms were an unmitigated disaster, but they'll never admit it publicly because it would mean having to admit that they were duped."

"These bishops, why are they like this?"

He quoted St. Athanasius: "The floor of Hell is paved with the skulls of bishops." After a brief pause, he added: "But they're not all bad. You said you're from Mississippi, right?"

"Yes."

"There's a bishop in Biloxi, a black man, who offers Latin Mass once a month in the Cathedral."

I couldn't believe my ears. "That's Bishop Joseph Howze! He confirmed me, and he also confirmed my wife." I had seen the flyer from the Coalition in Support of *Ecclesia Dei*, and the once-a-month Mass in the Biloxi Cathedral was one of very few locations anywhere in the South that one could find a Latin Mass. From then on, we attempted to time our visits home to Mississippi to correspond to the weekends that Bishop Howze would offer the traditional Latin Mass in the Cathedral.

The author being confirmed at St. Thomas Aquinas Catholic Church
in Hattiesburg, MS, by Bishop Joseph L. Howze, First Bishop of the
Diocese of Biloxi, in 1973. Bishop Howze was generous in giving
permission for the Latin Mass within his diocese.

"Why doesn't the pope do something about the situation?"
I asked Will.

"I don't think he sees the need. Remember Archbishop Lefebvre? Look what happened to him."

"But I thought you said he wasn't excommunicated for pro-
tecting the Latin Mass. You said it was for something else. What
exactly did happen?"

"It's a long story. It requires a bit of background."

CHAPTER 5

TIME BOMBS AND TROJAN HORSES

THE CATHOLIC PRESENCE IN THE AMERICAS PRE-dated the establishment of the United States by hundreds of years. The Royal and Pontifical University of Mexico was founded in 1551 by Charles I of Spain, and the National University of San Marcos in Lima, Peru, was founded the same year. When the thirteen English colonies broke from England in 1776 there were Catholics in every colony. Fr. John Carroll was appointed as the first bishop in 1789, with Baltimore as the seat of the first diocese. The religious freedom clause of the Bill of Rights of the U.S. Constitution provided no real obstacle to the eventual growth and flourishing of the Catholic Church in America. A constitution based on Natural Law was bound to ensure that those seeking their own eternal well-being would be unhampered in working out their salvation. Religious orders flourished, missionary work spread, and the Church quietly fulfilled Her mission in a secular state founded largely by Protestants. It was inevitable that there would be clashes between the two parts of society, the Church and the State, as each worked to fulfill its purpose. The former was for the salvation of souls, and the latter was for the regulation of temporal affairs such as the delivery of the mail, providing the common defense, and maintenance of infrastructure. The framers of the U.S. Constitution understood the principle of subsidiarity and incorporated it in their design, so that those roles not specifically enumerated for the federal government were left to state and local governments. For those Catholics serving in the military, there would be no military vicariate for a long time. Their pastoral care was entrusted to local ordinaries in whatever way best suited the situation.

Over the course of the next nearly two hundred years unparalleled contributions to the overall benefit of America came from U.S. Catholics. Religious orders founded schools and hospitals. Fr. Pieter-Jan De Smet, a Flemish (Belgian) Jesuit, carried out missionary work among the Native Americans so effectively that in 1868 he became an intermediary between Sitting Bull and the U.S. Government. There were countless other missionaries like him whose stories are not as well known. Sr. Blandina Segale, a twenty-two-year-old Sister of Charity, was sent into the Wild West to found a school. Along with thousands of others like her, she played a key role in the taming of that unruly region.[1] Saint Frances Cabrini, born in Italy, founded the Missionary Sisters of the Sacred Heart of Jesus, arriving in America in 1889 to serve Italian immigrants. Saint Elizabeth Ann Seton, the first saint to be born in America, founded the Daughters of Charity of Saint Vincent de Paul. Saint Katharine Mary Drexel founded the Sisters of the Blessed Sacrament and dedicated her life to working with minorities. The list of American saints, *beati*, venerables, and martyrs is impressive. All of this missionary work was done by Catholics who prayed in Latin during the Mass. The Mass that these heroic Catholics attended and prayed was the traditional Latin Mass that would eventually become a source of bitterness to so many contemporary "Vatican II" Catholics. It is also important to note that, until the time of World War I, there was still no military vicariate.

There are contemporary Catholics who will assert that the separation of Church and State and the placement of all religious creeds on an equal level were a departure from long-standing Catholic social teaching. That may be true, and it is also true that the proliferation of non-Catholic Christian sects was a departure from the traditionally understood unity that should exist among the followers of Christ. These divisions originated in Europe and were inherited by the newly founded republic. Also, unity did in fact exist among Catholics themselves, largely based on unchanged doctrine and a liturgy that they all prayed in common. Groups of immigrants coalesced together in various regions of the country, but, aside from sporadic communities of Eastern Catholics, the liturgy used by all

---

[1] Sr. Blandina Segale, *At the End of the Santa Fe Trail* (Arcadia Press, 2019); many editions of this work are available.

was codified by Pope St. Pius V in 1570. Hence it was possible for an Italian immigrant to attend Mass and pray with German or Irish immigrants and, apart from the language of the sermon, expect no substantive difference in the liturgy itself or even in how it was conducted. A Catholic could travel across the country or the world and expect to find the same doctrine and order of worship in any Catholic community. The Mass was always the same, and it was, with only minor differences, what we now call the traditional Latin Mass.

There is no question that the Catholic Church in America flourished under the provisions of the U.S. Constitution. This is not to say that life was ideal for American Catholics at any point. They were often at the mercy of hostile and very powerful financial forces and a deceitful press that misused First Amendment freedom to deceive and lead astray as many "Papists" as possible, or at least to ensure that they never held any powerful political positions. But how often were Catholics martyred over the course of the first two hundred years of American history? Aside from periodic slanders, economic suppression, and the occasional isolation of living among non-Catholics or pagans, one can point to very few real persecutions. In his *Democracy in America*, Alexis de Tocqueville noted some important distinctions between American and European Catholics. American Catholics were fervently devout and, despite living in a "democratic" society, they nevertheless held uniform beliefs and practices, commonly understood doctrine, and the very unifying Latin Mass. He saw Catholicism in America as compatible with the equality demanded among those who live under a representative political regime. He noted the difference between this perspective and the French perspective in which Catholicism was seen as a rival to democracy.

In 1917, the Apostolic See created the Military Vicariate and appointed Bishop Patrick Hayes, an Auxiliary Bishop of the Archdiocese of New York, as the first bishop. This was convenient, as New York was the main port for U.S. troops and the chaplains deploying with them. By the beginning of World War II (1939) there were about 21 million Catholics in the U.S. out of a total population of nearly 132 million, or about 16 percent. All, except a tiny minority of Eastern-rite Catholics, attended Mass in Latin as it had been codified after the Council of Trent. The estimated Catholic participation in the armed forces during World War II was much higher, somewhere between 25 and 30 percent. Some U.S. bishops went out of their way to assure the president and the non-Catholic public of the patriotism of U.S. Catholics, and Archbishop Francis Spellman of New York assured troops that "in serving your country in a just cause, you are also serving God."[2] Hennessey refers to this period as a rite of passage for Catholics in America. It would eventually increase their prestige at all levels of society. John F. Kennedy, the future president, commanded a patrol torpedo (PT) boat in the Pacific. He was injured and some of the crew were killed when the boat was rammed by a Japanese destroyer, and for his heroic efforts in saving the remaining crew over the next few days he was awarded the Navy and Marine Corps Medal.

Of the military chaplains that served in the U.S. Armed Forces during World War II, 3,036 were Catholic, and most of them were of the Roman rite. Commander Joseph T. O'Callahan, a Jesuit and a U.S. Navy chaplain, received the Medal of Honor for his actions on the U.S.S. Franklin after it was bombed in March 1945. Lt. Joseph Lafleur, a U.S. Army chaplain, died after having spent two years in Japanese prison camps. A cause for his beatification has been initiated by the Diocese of Lafayette, Louisiana.

Although Catholics served in all branches and from all parts of America, support for the war effort was not universal. Dorothy Day, who cofounded the Catholic Worker Movement, opposed U.S. participation in the war, but many members of this movement participated in the conflict in noncombatant roles. Out of nearly 12,000 conscientious objectors, only 137 were Catholic. All, including Dorothy Day, John F. Kennedy, the 3,036 Catholic chaplains, and the rest of the Catholic participants of the war, prayed in Latin and received the sacraments in a manner that would soon be driven underground. Within twenty years of the end of World War II, hundreds of thousands of surviving Catholic service personnel would not be allowed to pray according to the rite that had sustained them through the Normandy Invasion, Bastogne, the island-hopping struggle for the Pacific, and the postwar occupations. How did the

---

[2] James Hennessey, S.J., *American Catholics: A History of the Roman Catholic Community in the United States* (Oxford University Press, 1981), 280.

traditional Latin Mass that comforted these courageous veterans eventually become such a source of bitterness?

By the end of World War II there were numerous orders of religious, both sisters and brothers, running schools throughout the country. This had the effect of keeping education affordable and forcing the public schools to behave decently as they attempted to compete with the well-established network of Catholic schools. Religious orders ran hospitals, and with much of the labor done by people committed to vows of poverty, these hospitals kept the standard for medical care both affordable and moral. A made-in-America missionary order, the Maryknolls, had missionaries working all over the world. Some, including Bishop Xavier Ford, were martyred in Communist China. Bishop Ford was tortured and died in prison in 1952. A Catholic chaplain, U.S. Army Captain Emil Kapuan, was posthumously awarded the Medal of Honor for his actions before, during, and after his capture at the Battle of Unsan during the Korean War in November 1950.

In short, the Catholic Church in America was alive and well. The 1960s began peacefully, with the election of John F. Kennedy, the decorated veteran PT boat commander, as the first Catholic president. In the early days of the Second Vatican Council, held from 1962 to 1965, very few Catholics saw trouble brewing. We were told that there would be some changes; I was even made aware of this as a first-grade student in a Catholic school. American Catholics were obedient and trustful toward their priests and bishops. They trusted the pope.

The few people who had the insight to foresee the pending disaster held several common attributes. They were deeply devout, holy, and unusually intelligent. The error behind the looming calamity was philosophical, and detecting philosophical errors requires either a brilliant philosopher or a mystic guided by holy insight. No outright heresy was to be pronounced at this council, yet it was flawed from the start because it was based on a false premise: that man had somehow changed, and this new "modern man" required a different ecclesiastical approach. Dietrich von Hildebrand, a German philosopher referred to by Pope Pius XII as a "twentieth-century Doctor of the Church," pointed out:

> We wish to repeat emphatically: there is no closed, homogeneous epoch in history; there is no "modern man." And most important of all, man always remains the same in his essential structure, in his destiny, in his potentialities, in his desires and moral dangers; and this is true notwithstanding all the changes that take place in the external conditions of his life.[3]

This false premise, the myth of a "modern man," was the Trojan horse. From this false premise, the following errors flowed, crippling the council from the start:

> 1. That this imaginary "modern man" had something new to offer, and thus needed a new pastoral approach.

> 2. This approach demanded that this "modern man" should somehow be included in the governance of the church, and that the liturgy should be modified to accompany him.

> 3. That Communism, under which nearly half the world's population was enslaved, was not to be confronted outright.

As a precondition for Russian Orthodox clergy to sit at the table, the Metz Pact was signed. According to this agreement, Russian Orthodox clergy would participate in the council only under the condition that Communism would not be expressly criticized.[4] The incompatibility of Communism and Catholic doctrine had been clarified as recently as 1958 by Pope Pius XII in his encyclical *Apostolorum Principis*. It would have been essential for the council to emphasize this incompatibility as global communism threatened the entire world. The Council's very opening in October 1962 coincided with the Cuban Missile Crisis, during which the world came closer than it ever had to nuclear annihilation.

---

[3] Dietrich von Hildebrand, *Trojan Horse in the City of God* (Franciscan Herald Press, 1967), 132.

[4] Romano Amerio, *Iota Unum: A Study of Changes in the Catholic Church in the Twentieth Century*, trans. John P. Parsons (Sarto House, 1996), 75–76. For an excellent summary of the evidence, see Edward Pentin, "Why Did Vatican II Ignore Communism?," *The Catholic World Report*, December 10, 2012, www.catholicworldreport.com/2012/12/10/why-did-vatican-ii-ignore-communism/; see, as well, Matthew Cullinan Hoffman, "Revisiting Vatican II's lost condemnations of communism after Pope Francis encourages Marxists," *LifeSiteNews*, January 11, 2024, www.lifesitenews.com/news/vatican-iis-lost-condemnations-of-communism-revealed-to-public-for-first-ti.

Most of the bishops assembled in Rome for the Second Vatican Council had no revolutionary agenda in mind. They were assembled to agree upon modest, pastoral and disciplinary changes that could help the Church to grow, evangelize, restore Christian unity, and more fully accomplish Her mission on Earth, which is to bring to all of mankind the means of salvation. The Council defined no new dogmas, as it was intended to be a pastoral council. This fact, still poorly understood, was confirmed by three popes. John XXIII stated at the start of the council: "The substance of the ancient doctrine of the deposit of faith is one thing, and the formulation in which it is clothed is another. And it is the latter that must be taken into great account, with patience if necessary, measuring everything by the forms and proportions of a magisterium primarily pastoral in character."[5] Paul VI stated at the end: "The magisterium of the Church did not wish to pronounce itself under the form of extraordinary dogmatic pronouncements."[6] The future Benedict XVI confirmed this point in an address to the Chilean bishops: "The truth is that this particular Council [Vatican II] defined no dogma at all, and deliberately chose to remain on a modest level, as a merely pastoral council."[7]

Some disciplinary changes were proposed. Modification would be made to the *Missale Romanum*, but the modifications were to be minor. Of course, the prayers would remain in Latin; any radical change was unthinkable. It was all to be just a bit of modernization for the "modern man." Most importantly, tradition was to be safeguarded! Several articles from *Sacrosanctum Concilium*, the document on the liturgy that was to come from Vatican II, made that point clear. In accordance with Article 4, all rites were to be held as being of equal right and dignity, and they were to be preserved and fostered. According to Article 23 there were to be no innovations unless the good of the Church required them. According to Article 36, the use of Latin was to be preserved in the Latin rites, and according to Article 116, Gregorian chant was to be given "pride of place" (*principem locum*, which literally

---

[5] *Gaudet Mater Ecclesia*, October 11, 1962, in the translation provided by Joseph A. Komonchak, https://jakomonchak.wordpress.com/wp-content/uploads/2012/10/john-xxiii-opening-speech.pdf.
[6] Discourse at the closing of Vatican II, December 7, 1965.
[7] Joseph Cardinal Ratzinger, future Pope Benedict XVI, in an address to the Chilean bishops, July 13, 1988.

means "chief/foremost place") in liturgical services. According to Article 54, steps were to be taken so that people would be able to pray and sing in the language of the Latin rite during the Mass.

Although most bishops had no revolutionary inclinations, a few in favor of radical change saw their opportunity and took control of the agenda. Recognizing that most bishops would not agree with their agenda, they maneuvered to put forward a set of documents that would be agreed upon by the majority, but would be vague enough that the progressives could carry out their agenda in the following years. The draft documents carefully prepared before the council were abandoned[8] while ambiguous texts were advanced, leaving loopholes and vulnerabilities. Being radicals, these men believed the ends justified the means. Under the direction of Archbishop Annibale Bugnini, who controlled the committee responsible for the implementation of the liturgical reform, they laid the groundwork for a complete overhaul of the liturgy, gaining apparent approval by most bishops without raising much suspicion. Most bishops were operating in good faith, so there was little resistance.

It was after the end of the Council that the time bombs began to detonate. Reviewing some of the articles of *Sacrosanctum Concilium* again, the problem becomes obvious. The time bombs are inserted with appropriate warning:

> ARTICLE 4. In faithful obedience to tradition, the sacred Council declares that holy Mother Church holds all lawfully acknowledged rites to be of equal right and dignity; that she wishes to preserve them in the future and to foster them in every way. [Time bomb warning!] The Council *also desires* that, where necessary, the rites be revised carefully in the light of sound tradition, and that they be given new vigor to meet the circumstances and needs of modern times.

This clause provided the necessary loophole for any kind of revision, even the most radical. Who decides what is in the "light of sound tradition" and who decides what meets the "circumstances and needs of modern times?" Prayers at the Foot of the Altar were eliminated, the Mass readings were distributed over two- and three-year rotation periods, the Eucharistic prayers now

---

[8] Except for the draft of *Sacrosanctum Concilium*, of which Annibale Bugnini had been the compiler. This draft survived the progressivist purge.

came in multiple options, and other unprecedented revisions were enacted which were neither careful nor in light of any tradition. All lawfully acknowledged rites having equal right and dignity was clearly not applied to the timeless Tridentine rite that Catholics were accustomed to at the time. No provision was in place for its continued use, although as we would eventually find out, it was never formally abrogated or repealed.

> ARTICLE 23. That sound tradition may be retained [time bomb warning!] *and yet* the way remain open to legitimate progress, careful investigation is always to be made into each part of the liturgy which is to be revised.

Who decides what is "legitimate progress?" Some faction that believed that Latin and Gregorian chant should be abolished? This paragraph concluded with:

> Finally, there must be no innovations unless the good of the Church genuinely and certainly requires them; and care must be taken that any new forms adopted should in some way grow organically from forms already existing.

Given the results, is there any good reason to think that the "good of the Church genuinely and certainly" required any of the changes that were enacted?

> ARTICLE 36. Particular law remaining in force, the use of the Latin language is to be preserved in the Latin rites. [Time bomb warning!] But since the use of the mother tongue, whether in the Mass, the administration of the sacraments, or other parts of the liturgy, frequently may be of great advantage to the people, the limits of its employment may be extended.

This "extension" of the limits of the use of the vernacular went to the most radical level possible. The Latin language was not preserved at all, but rather was abruptly and completely replaced by the vernacular. This barbaric act was accompanied by a media montage of slogans proclaiming that "Vatican II changed the Mass from Latin to the vernacular." A chorus of media outlets advanced this narrative, championed it, and to this day they continue propagating this lie. For English-speaking Catholics, that meant a fundamentally flawed English translation that remained uncorrected for almost forty years until 2011, during the pontificate of Pope Benedict XVI.

ARTICLE 54. [Here the time bomb comes up front:] In Masses which are celebrated with the people, a suitable place may be allotted to their mother tongue. This is to apply in the first place to the readings and "the common prayer," but also, as local conditions may warrant, to those parts which pertain to the people, according to the norm laid down in Art. 36 of this Constitution. [Now back to an attempt at reassurance:] Nevertheless steps should be taken so that the faithful may also be able to say or to sing together in Latin those parts of the Ordinary of the Mass which pertain to them.

And wherever a more extended use of the mother tongue within the Mass appears desirable, the regulation laid down in Art. 40 of this Constitution is to be observed.

The reference to Article 40 is important. It provides for taking the local culture into consideration, to discern the circumstances under which "a more extended use of the mother tongue" was needed. As it turned out, this "more extended use of the mother tongue" was needed throughout the world, within every civilization and in every culture. The part about being able to pray or sing together in Latin was rendered meaningless.

ARTICLE 114. The treasure of sacred music is to be preserved and fostered with great care. Choirs must be diligently promoted, especially in cathedral churches, [Time bomb warning!] but bishops and other pastors of souls must be at pains to ensure that, whenever the sacred action is to be celebrated with song, the whole body of the faithful may be able to contribute that active participation which is rightly theirs....

Here it was understood, apparently, that if a single person in the congregation could not sing a hymn or chant that was in Latin, it must be replaced. The actual implementation brought the sudden introduction of guitars and artificially concocted folk music.

ARTICLE 116. The Church acknowledges Gregorian chant as specially suited to the Roman liturgy: therefore, other things being equal, it should be given pride of place in liturgical services. [Time bomb alert!] But other kinds of sacred music, especially polyphony, are by no means excluded from liturgical celebrations....

This very clear guidance was "implemented" in a way that completely eradicated Gregorian chant. Polyphony was also excluded, even though specifically mentioned as being worthy of preservation. The reason is obvious: its existence in the Mass would be a violation of Article 114 which required that "the whole body of the faithful may be able to contribute that active participation which is rightly theirs." In other words, the music had to be something that everyone in the congregation could sing along with. Obviously, that doesn't include polyphony.

In 1969, four years after the council closed, a completely new rite was introduced. Its introduction bewildered and confused Catholics, who were led to believe that this new rite was the natural outcome of Vatican II's document on the liturgy, *Sacrosanctum Concilium.* Bishops throughout the world, including those operating in good faith, were also led by a chorus of media outlets to believe that these radical changes were what the laity — or the mythical "modern man" — demanded. Although this radical revision was not what the bishops had agreed upon, they had in fact tacitly signed off on the ambiguous document.

What happened was nothing less than revolution. Fr. Joseph Gelineau, S.J., a liturgical expert at the French National Pastoral and Liturgical Center, who was considered by Archbishop Bugnini to be one of the great masters of the international liturgical world, stated it bluntly:

> Let those who like myself have known and sung a Latin-Gregorian High Mass remember it if they can. Let them compare it with the Mass that we now have. Not only the words, the melodies, and some of the gestures are different. To tell the truth, it is a different liturgy of the Mass. This needs to be said without ambiguity: the Roman rite as we knew it no longer exists. It has been destroyed.[9]

His use of the word "destroyed" gives us some insight into whether or not he thought that what he was doing was for the good of the Church. This honest appraisal also obliterates the notion that the new Mass somehow grew organically from an existing form. Fr. Yves Congar, one of the artisans of the "reform," said as much: "The Church has had, peacefully, its October

---

[9] *Demain la liturgie: Essai sur l'évolution des assemblées chrétiennes* (Cerf, 1976), 9–10. Gelineau continues: "Some walls of the former edifice have fallen while others have changed their appearance, to the extent that it appears today either as a ruin or the partial substructure of a different building." Quoted by Michael Davies, *Pope Paul's New Mass* (Angelus Press, 2009), 82.

Revolution."[10] This quotation is disturbing, since Fr. Congar is mentioned by Pope John Paul II as a priest with whom he had the "good fortune to work" and to whom he was "particularly indebted."[11]

It may be difficult for those who did not live through this "peaceful" revolution to understand that it was indeed a revolution, but that's what it was. Accompanying this revolution in the liturgy was the growth of a radical, immoral, and angry Catholic social activism. Void of a good and holy spiritual element, this movement sought to topple governments, realign alliances, redraw borders, place women in seminaries, and rewrite legislation. It was typically, but not always, aligned with the political left. When these Catholics found themselves advancing ideas that ran counter to Catholic moral teachings, they would side with their political allies and simply ignore Catholic morality. Some of the more extreme spinoff movements of this era included Revolution Theology and pro-choice Catholicism. Homosexual activism would come later.

The devout Catholic witnessing this horrifying decay of Catholic culture, morality, and decency had not only the spectacle of a dying civilization to cope with Monday through Saturday, but then on Sunday morning he had to face bad liturgy and guitar music. Gone were the beautiful Catholic hymns, so meaningful and timeless. Out with *O Sanctissima* and in with "Sons of God." Out with *Panis Angelicus* and in with Kumbaya. (Yes, we sang this, for those too young to remember. And yes, we held hands while doing so.) Thomas Day describes what he calls the triumph of bad taste in *Why Catholics Can't Sing*. A holder of a Ph.D. in musicology from Columbia University, a member of the American Guild of Organists, and a man of exceeding good humor, he describes a scene that took place sometime after the revolution. He was in the church practicing the organ while a Cuban refugee who spoke no English mopped the floor. While playing, he drifted into the melody of the *Veni Creator Spiritus*. Dr. Day describes the encounter:

> Suddenly, the Cuban man dropped his mop and came
> dashing up the stairs of the organ loft. When he got to
> the organ (a little out of breath) he started singing the
> grand old hymn, which he had learned as a youngster.

---

[10] Yves Congar, O.P., *Le Concile au jour le jour deuxieme session* (Cerf, 1964), 115.
[11] John Paul II, *Crossing the Threshold of Hope* (Alfred A. Knopf, 1994), 159.

> For a few brief minutes we were united by a Latin hymn
> dealing with theological complications we could barely
> follow. What united us was the sound of something
> uncommonly beautiful, something which did not come
> from Cuba or the United States, but from the "highest
> common denominator."[12]

It is helpful to ponder this example, because there is another false narrative that asserts that the preconciliar Mass was only for rich people, for the elites, for noblemen; the upper crust. In reality, it was the Mass for *all* Catholics of the West, and regardless of social status any devout Catholic loved the Church's beautiful traditions, especially its musical tradition. Why would priests and bishops want to suppress the beautiful Catholic music tradition in favor of *Glory and Praise* (1984), which, according to Dr. Day, "makes almost a complete break with the past." With few exceptions "there is no old music in this collection, nothing written before the 1960s. The past is repudiated."[13] Why, with all the talk of "the poor" and "social justice," could the poor Cuban janitor not have the beauty of the Catholic music tradition during Mass rather than only while mopping an empty church? Such music was banished to choir performances at colleges, universities, and high schools with a good choral program.

What was a devout Catholic to do under these circumstances? At some point any person who loves the Church reaches a limit. If the ideological agenda doesn't offend, the music does. If not the music, the ugly architecture, the felt banners, the banal and lazy vernacular, the sheer lack of piety and the disregard for anything spiritual, most especially silence. What happened to the belief in the Eucharist? If people still believed it was the real Body and Blood of Jesus, why were they suddenly touching the consecrated hosts with unconsecrated hands? Why did so many women no longer cover their heads in respect for the Real Presence? Why were men wearing jogging suits to Mass?

The generation of Americans that had been young adults during World War II is often referred to as the "greatest generation." After their heroic participation in an epic struggle,

---

[12] Thomas Day, *Why Catholics Can't Sing: The Culture of Catholicism and the Triumph of Bad Taste* (Crossroad, 1990), 105. Oddly, this anecdote is missing from the second revised edition of 2013.

[13] Day, *Why Catholics Can't Sing*, 70.

this same generation of Catholics laid down for the Vatican
II steamroller. Why did they allow the spiritual patrimony on
which their civilization was built to be discarded? Even the most
intelligent and gifted surrendered, threw down their arms, and
let the enemy occupy their civilization so soon after they had
successfully completed their "rite of passage."

The degree to which Catholics recognized these problems
depended on how well they knew their faith. Those gifted with
exceptional intellect or supernatural spiritual insight became
alarmed as events unfolded; Dietrich von Hildebrand was one
of the first to sound the alarm. Similar concerns were expressed
by Evelyn Waugh, J.R.R. Tolkien, Fr. Gomar DePauw, Fr. Malachi
Martin, John Senior, and many others, but they were up against
a force with which they were unable to compete. While the most
astute recognized trouble immediately, for others it took much
longer. Some of us were just beginning to figure it out in the
early 1990s. It was difficult to avoid the conclusion that what
was happening in the Church did not come about because of
something from within the Church, but rather because of an
enemy hostile to the Church. Pope Paul VI spoke of the "smoke
of Satan" in the sanctuary: "From some fissure the smoke of
Satan entered into the temple of God."[14] He also spoke of the
"autodemolition" of the Church. The decline in baptisms, the clo-
sure of schools, the dismantling of religious orders, the unhab-
iting of nuns, and the collapse of priestly vocations stunned
Catholics in the pews. The better one knew his faith, the more
painful Mass attendance became. Mass attendance declined by
millions. A quick view of the numbers of seminarians entering
seminaries in the United States showed that a vibrant, healthy,
and increasing number of entries up to 1965 had been replaced
by an exponential decay.[15] Nature does not allow reversals
of exponential decay. The seminaries in the U.S. would never
again meet pre-Vatican II levels. Millions of souls would be left
without the sacraments, and lose their faith or never discover
it. The Church in America was in decline.

The U.S. Armed Forces reflect U.S. society; therefore, it was
natural that along with the post-conciliar decline in the Catholic

---

[14] Pope Paul VI, June 29, 1972, on the occasion of the ninth anniversary of his
coronation.
[15] See references in note 4 on page 33.

priesthood came a decline in the number of Catholic chaplains. There was nothing about the 1984 establishment of the Archdiocese for the Military Services that would somehow alleviate this national trend. Along with the slow decline in Catholic schools, Catholic religious orders, and Catholic chaplains in the military came a decline in Catholic moral thinking among military men. With the lifting of the "rules" that governed the behavior of Catholics came the gradual—and very dangerous—loss of regard for human life and the loosening of restrictions regarding the justified use of military force.

CHAPTER 6

GIRLS WILL BE BOYS

For there will come a time when they will not endure the sound doctrine; but having itching ears, will heap up to themselves teachers according to their own lusts and they will turn away their hearing from the truth and turn aside rather to fables.[1]

Roman Catholic Christianity has a problem with women. This problem is deeply rooted in its history, in its assumptions about gender and sexuality. The foundational thinker of Latin Christianity, St. Augustine, in the late fourth and early fifth centuries established certain assumptions that still plague Catholicism.[2]

MY HEAD WAS SPINNING AS I CONTINUED TO unpeel the layers of dishonesty and confusion over the status of the traditional Latin Mass that Lorri and I had come to love in such a short period of time. This period of intense research ended abruptly with an inevitable military move. The Army was enacting some serious post-cold-war manpower reductions, in some cases literally buying members out of their service by offering huge departure bonuses. Amid reductions, the AFIT program I was part of was cut and the four of us Army officers were all reassigned.

During the summer of 1993 I reported to the U.S. Army Computer Science School at Fort Gordon, Georgia. Lorri was expecting another baby, and there was a four-month wait for base housing. We rented an apartment in Augusta, expecting to be offered base housing after the baby was born. After spending a few days unpacking, arranging furniture, hanging pictures, stocking the refrigerator, and organizing the small apartment we were finally settled in. As I hung the last picture the phone rang. It was the Housing Office. A house on base was now available for us! We

---

[1] 2 Tim. 4:3-4.

[2] Rosemary Radford Ruether, "Women, Reproductive Rights and the Catholic Church," *Feminist Theology*, vol. 16, no. 2 (January 2008): 184-94.

spent the next few days repacking everything in the apartment to move three miles away to Fort Gordon.

The big question hovering over us after this recent spiritual awakening was where to attend Mass. I scanned the list of locations for a Latin Mass approved by the local bishop. There were none on Sundays anywhere in Georgia! One Thursday a month there was a Latin Mass in Atlanta, at 7:00 p.m. Fat chance I'd ever be able to make it to that. Whatever happened to the pope's plea for generosity with the old rite? We decided that we would look around for a good parish just as we had done in the past.

There was a weekly Mass at the Fort Gordon post chapel, so we decided to go there first. There it was again! Girls were dressed up as altar boys and serving at the altar. At that time this was a serious abuse of the liturgy. When the Church has liturgical norms, they are to be respected, and this particular innovation was something that had been explicitly forbidden.[3] To see this casual disregard for the liturgical norms of the Church on a military installation, with the involvement of a Catholic chaplain, was disturbing. Apparently it was more important to pander to feminists than to worry about what the Church had to say.

What was becoming obvious to me by now was this: The old rite, which the Holy Father himself had asked the bishops to allow generously, was forbidden for all intents and purposes. At the same time, this abuse of the liturgy, which really was forbidden by the Church, was in practice everywhere. The Church in the U.S. had been hijacked by the feminist agenda.

I met with the chaplain, Fr. David Kernighan, to discuss it. He listened politely to my concern, then laughed, waved his hand, and expressed amazement that I would have second thoughts about such a trivial issue. According to Fr. David, there was no need to respect such guidance from the Church. The pope didn't understand how things were in America; he had a medieval mentality, as did Mother Theresa and other old people. Surely, I didn't want to be like them! When he said the word medieval, the 'e' was stretched out as long as possible.

Medieeeeeeeeeval!

Seeing that this didn't impress me, he dismissed me as quickly as possible.

---

[3] See the excerpts from *Liturgicae instaurationes* (1970) and *Inaestimabile donum* (1980) quoted above on page 36.

We found a nice, rather traditional parish in downtown Augusta and began attending Mass there. The priest gave good sermons, and he was obviously very orthodox. He took an active interest in his flock and came to our house for dinner when we invited him. We missed the Tridentine Mass, but since we knew that we were only here temporarily we were content with the situation. We would try to be happy Novus Ordo Catholics with a great appreciation for the Latin Mass.

On New Years Day 1994 our second child, William, was born. It was about this time that I discovered that we would be transferred in six months. Life in the Army can involve lots of moves, and we were getting our share of it.

Then, out of the clear blue, something bizarre happened. In March of that year a letter was issued from the Pontifical Council for the Interpretation of Legislative Texts. I say "issued," but it's not clear what exactly happened. A letter from this office, bearing no signature, was sent by fax to news agencies all over the world proclaiming a new truth. For some bishops, the first they would hear of it was when they read about it in the newspaper.[4] The letter stated, contrary to previous instructions, that now there was no clear prohibition of female altar servers because it was not mentioned in the new Code of Canon Law.

Well! In that case it's also not forbidden to begin Mass with a motorcycle procession. It's not mentioned in the new Code of Canon Law, so there is no clear prohibition. Nor is a skateboard procession prohibited.

The highest authorities in the Church had decided, in contradiction with previous instructions directly from popes, to allow female altar servers. In other words, what had been wrong up until now was suddenly not wrong. At least the letter also made provisions that we would not have to accept this new practice:

> The permission given in this regard by some Bishops can in no way be considered as binding on other Bishops. In fact, it is the competence of each Bishop, in his diocese, after hearing the opinion of the Episcopal Conference, to make a prudential judgment on what to do, with a view to the ordered development of liturgical life in his own diocese.[5]

---

[4] Bishop Howze told me personally that he first heard about it when he read it in the local newspaper.

[5] Congregation for Divine Worship and Discipline of the Sacraments, Communication sent to the presidents of episcopal conferences, March 15, 1994.

And:

> At the same time, however, the Holy See wishes to recall
> that it will always be very appropriate to follow the
> noble tradition of having boys serve at the altar. As is
> well known, this has led to a reassuring development
> of priestly vocations. *Thus, the obligation to support such
> groups of altar boys will always continue.*

The only way to support groups of altar boys is, obviously,
to have exactly that: groups of altar boys. But somehow, I sus-
pected that U.S. bishops would find a way to mandate the new
permission whether we liked it or not. Sure enough, they quickly
issued new guidelines. In June of that year, they held a special
assembly to discuss implementation. The result included the
following suggested points for developing diocesan guidelines:

> 1. Although institution into the ministry of acolyte is
> reserved to lay men, the diocesan bishop may permit the
> liturgical functions of the instituted acolyte to be carried
> out by altar servers, men and women, boys and girls.
> Such persons may carry out all the functions listed in no.
> 68, para. 2 and nos. 142–147 of the General Instruction of
> the Roman Missal. The determination that women and
> girls may function as servers in the liturgy should be
> made by the bishop on the diocesan level so that there
> might be a uniform diocesan policy.
>
> 2. No distinction should be made between the functions
> carried out in the sanctuary by men and boys and those
> carried out by women and girls. The term altar boys
> should be replaced by servers. The term server should
> be used for those who carry out the functions of the
> instituted acolyte.[6]

Had the original document from the Vatican said anything
about replacing the term "altar boys" with "servers"? Of course
not. And of course, the assembly made no mention of the "noble
tradition of having boys serve at the altar" from the original
document, or the obligation to support groups of altar boys.

In less than six months what had previously been a grave
abuse of the liturgy was now mandatory in most parishes

---

[6] Guidelines prepared by the Committee on the Liturgy and presented to the
National Conference of Catholic Bishops for discussion at the June 1994 Special
Assembly, Thursday, June 16, 1994.

throughout the U.S. But how can something be a grave abuse of the liturgy one day and then become mandatory the next day? How is this even possible?

The answer is simple. It's not. What was given was permission, not a mandate. Anyone can see that by reading the plain text: *may permit*. Yes, of course, it was to be left to the local bishop to decide the policy on the diocesan level, but how many would have the courage to stand up against the feminist spirit that was driving this capitulation? As it turned out, there were only two such bishops: one in the diocese of Arlington, Virginia and another in the Diocese of Lincoln, Nebraska.[7] Everywhere else, not only did it become diocesan policy, but it also became mandatory. It would be many years before the Holy See would clarify that this was not something that any bishop has the authority to force on anyone, and that every priest has the right to have exclusively boys serving at the altar.[8] Another fact that has remained hidden throughout the years since 1994 was that this practice was never permitted in the pope's own Diocese of Rome. But this secret was not revealed to U.S. Catholics, who were led by their bishops and priests to believe that it was something the Holy Father was demanding.

During his long pontificate Pope John Paul II addressed this issue only once, in 1980, and he condemned it. On the other hand, he addressed the need to retain Latin in the liturgy and allow the 1962 Missal on multiple occasions. Yet, within the United States it was the reverse: the use of Latin in the liturgy was highly restricted at best, and suppressed in most cases, and the

---

[7] The Arlington diocese finally capitulated to the Zeitgeist when, in March 2006, Bishop Paul Loverde broadly allowed female altar servers for the first time. As of the writing of this note (2026), Lincoln remains the only diocese in the USA — out of 178 Latin-rite dioceses — that has a diocesan-wide policy of only male altar servers. As it happens, the diocese of Lincoln inaugurated a new diocesan seminary in 1998, St. Gregory the Great, and welcomed the building of Our Lady of Guadalupe Seminary for the Priestly Fraternity of Saint Peter, inaugurated in 2000. Both are flourishing.

[8] In July 2001, the Congregation for Divine Worship issued a response to a bishop's question (*dubium*) concerning the possible admission of girls and women as altar servers. The response made it clear that only a diocesan bishop may decide whether to permit female servers in his diocese; furthermore, that no priest is obliged to have female servers, even in dioceses where this is permitted. The letter stressed that no one has a right to serve at the altar, and strongly reaffirmed that altar boys should be encouraged. Congregation for Divine Worship, *Notitiae* 421-22, vol. 37, nos. 8-9 (2001): 397-99.

feminization of the liturgy was forced practically everywhere on unsuspecting Catholics. Had the U.S. bishops responded according to how much attention the pope gave to both issues, there would have been Latin Masses throughout the land, and in a handful of radical feminist churches on college campuses there would be girls pretending to be altar boys.

As we have seen, the decision was to be left at the diocesan level so that each bishop could decide whether to allow this new practice. But it was soon clear that this was not something that most of them saw as a problem; they saw it rather as something to embrace, or to forcefully impose with threats, if necessary. Or perhaps, they lacked the courage to say "no, we won't be engaging in this novelty." In most cases the new policy was implemented with an iron fist. There was no escape from it. The ordinary diocesan priest who didn't want to go along with it was intimidated, threatened, and cajoled into compliance. Even at our parish in downtown Augusta, it appeared that there would be no avoiding it. The bishop had spoken. What else was there to do?

Our poor parish tried to organize a sodality of some sort that would minimize the damage while meeting this new requirement, but we would never see it implemented because I received orders for a new assignment and we would move before it came to pass.

But what about our future homes and parishes as the Army continued to move us around? Throughout the country the fruits of years of disobedience were paying off for the revolutionaries. Today girl altar boys, tomorrow women deacons, eventually women priests, bishops, and one day perhaps a woman pope! They truly believed that this was the only way that they could force the Catholic Church to accept abortion as a morally legitimate "choice." They would need a woman pope for that, and they were determined to get it.

So it was that in 1994, an act of rebellion, born out of the sexual distortions of post-1960s American liberalism, was suddenly mandated in Catholic churches throughout the land. What were we to do? Just live with this bizarre liturgical novelty as an act of obedience, knowing that it would cause grave harm to the future vocations of our children which we were bound to protect? Or were we to take some action to protect our families?

We are never obliged to obey in the case of sin or to participate in anything that would destroy our faith and risk our eternal salvation. If girls on the altar had been wrong until now, it was still wrong. Maybe it had been permitted to avoid having the American Church go into schism (as some opined), but we did not have to accept it.

We had one last hope for avoiding the constant barrage of propaganda and the loss of the faith in our family: the Holy Father himself had authorized the old Mass (an authorization that wasn't technically necessary since it had never been unauthorized to begin with). There would be no such silly problems to concern ourselves with if we assisted at the Mass of the Ages—if we could find it. The significance of the *Ecclesia Dei* document, with its plea from the pope for "wide and generous" availability of the old Mass, unaltered by vile political modifications, took on a new significance. This Mass would be our only way to avoid having faith-damaging and un-Christian politics become a part of our life in the Church. I was quite aware that there would be no end to the radicalization of Catholicism in the U.S. If the dissidents were so bold as to change God's Holy Scriptures and to defy the Church until they had their way, certainly they would continue in their revolt. I would have no part of it. We would not support this sexually confused rebellion against God's law, Sacred Tradition, and common sense. We would remain loyal to the Church, and we would do so by simply attending, from here on out, the Mass of the Ages. The Mass that the Saints had attended during their lifetimes. The Mass that the Modernists had not been successful in obliterating from the face of the earth.

The question was, how do you go about getting a Latin Mass at your local church?

CHAPTER 7

MONTGOMERY, ALABAMA

SOMETIME AFTER SERVING BETWEEN TWELVE and sixteen years as an officer, if you do your job reasonably well, you may have the opportunity to go to "Command and General Staff College." That is the term for the Army's school at Fort Leavenworth, Kansas, but each of the services has a center for education at the field grade level. The Navy has a command and staff college located in Rhode Island and the Air Force version is in Montgomery, Alabama. It's not unusual to attend the service school of a sister service. I received orders to report to the Air Command and Staff College in Montgomery for a year to study among, mostly, Air Force officers. There were also a few other officers from sister services and from allied nations.

I enjoyed the academics from the very first day. We read the great military classics, solved logistics problems, discussed military history, and made numerous presentations on a variety of topics. My background and computer skills gave me the dubious honor of showing fellow officers how to use their computers, the World Wide Web, and various software packages.

Lorri and I found a house to rent in a small town a few miles from Montgomery, with a Catholic Church nearby. The first Sunday we were there, a letter was read from the pulpit announcing that the new policy regarding "female altar servers" would be implemented in two weeks. They were looking for girl volunteers to serve. I looked at the faces of the boys on the altar for some sign of emotion. I saw none, but a few weeks later I noticed that none of them were even at Mass. I prayed for God to help my family through the difficult days that the Church was going through, and to show me the way to avoid participating in what I considered to be a desecration of the liturgy.

Birmingham was a little over an hour away, and there we could find three reasonable choices. At two different churches the traditional Latin Mass was offered on alternating Sundays. Also, there was Eternal Word Television Network (EWTN) in Irondale, outside of Birmingham. Irondale was, at the time, the home of the Poor Clares of Perpetual Adoration, founded by Mother Angelica, who was also the founder of EWTN. The new liturgy was followed there, but we knew that the feisty Mother Angelica would never allow such novelties as had been forced on unsuspecting Catholics practically everywhere else.

When my wife was young and still had no marriage plans, for a time she had a serious desire to become a nun. She contacted an order of women religious in Saint Louis and planned a visit. Upon her arrival she was dismayed to see that these women were wearing blue jeans and t-shirts, and they looked and behaved no differently than any other group of women who happen to be living together. "We work among the poor. If we are going to work among the poor, we should dress like them," they told her. She cut her visit short, returned home, and thankfully she eventually married me. The women religious, so helpful in the education and guidance of so many young Americans throughout the country until Vatican II, had by the 1990s all but ceased to exist. Not so for Mother Angelica and the sisters at the monastery in Irondale. They were young and vibrant, not dying of old age in blue jeans and t-shirts like so many who had bought into the deception.

We made trips as often as possible to Birmingham. But with my wife expecting again, I knew it would not be possible to make the drive every Sunday, so we tried to remain active within our parish. Soon after our arrival in July 1994, I was asked by the Confraternity of Christian Doctrine (CCD) coordinator if I would volunteer to teach, and I accepted. CCD is an acronym used in the modern age to avoid the antiquated word "catechism." She wanted me to teach the seventh-grade class, although she knew nothing about me. On Wednesday evenings from September through December I would meet with about ten students. I was given some very weak material to work with. It consisted of booklets that were void of any serious Catholic content or catechesis, but the pages were glossy and there were lots of flowers. However pretty it was, there was no way for me to use it for its intended purpose, so I decided that I would purchase some copies of the *Baltimore Catechism* and pay for them myself.

During one of the first meetings of the CCD teachers, we were asked for comments, or questions on what we were to do with the students. I said that I'd like to bring the students to the Tridentine Mass, and I asked whether there was a bus or a van available for field trips. The priest, Fr. Charles Troncale, became agitated and asked me to step into the hallway.

"What are you trying to do? Who do you think you are coming in here, trying to … what are you doing anyway?"

Now, I had never been yelled at by a priest before, but then again, I had never asked for a van to bring young people to the old Mass. His reaction left me speechless, but I gave him the benefit of the doubt, assumed that he was confused, and made an appointment to see him a few days later. I brought him some copies of documents from the Coalition in Support of *Ecclesia Dei*[1] and showed him the list which included the Mass times and locations in Birmingham. He apologized for his outburst and revealed that there was a group of Catholics in Montgomery holding Mass in the old rite without permission of the archbishop. He said that they were "crazy," and that he thought I was one of them. They had a chapel of some sort—it was normally a dance hall or an activity center, but they used it as a chapel one Sunday a month, "Our Lady of Lourdes Roman Catholic Chapel," and a priest from Cullman, north of Birmingham, came to offer Mass, without Archbishop Lipscomb's permission. Montgomery was in the Archdiocese of Mobile, where the only authorized Latin Mass was once a month near Mobile, more than two hours' drive away. I asked whether the archbishop made any effort to establish an authorized Mass in Montgomery. Fr. Troncale didn't know.

I decided to initiate a petition, and Fr. Troncale, to his credit, helped me. In October of 1994, I wrote a letter to Archbishop Oscar Lipscomb, of the Archdiocese of Mobile, in which I quoted

---

[1] The Coalition in Support of *Ecclesia Dei* was led for decades by Mary Kraychy, who died in 2025 at the age of 94. For a long time it was the only source of detailed and accurate information about where the traditional Mass was being celebrated in the United States. Its famous "red booklet" for the Mass has been distributed in the hundreds of thousands of copies, joined later by Requiem and Wedding booklets and by a Latin-Spanish booklet. Although moribund today, in its heyday the Coalition was vital in promoting the implementation, first, of John Paul II's *Ecclesia Dei*, and, later, Benedict XVI's *Summorum Pontificum*.

some key sentences from *Ecclesia Dei* and I asked that the old Mass be made available to those of us residing in the vicinity of Montgomery. Fr. Troncale even assisted me in putting the word out so that we could gather signatures for our petition. I think he felt guilty about the way he had treated me initially. Also, to Fr. Troncale's credit, when the CCD coordinator found out we were using the *Baltimore Catechism* and tried to demand that we stop, he told her, instead, to reimburse me for what I had spent.

With Fr. Charles Troncale, while attending Air Command and Staff College in Montgomery, Alabama. Fr. Troncale assisted us in circulating a petition for the Traditional Latin Mass in the Archdiocese of Mobile.

When we received Archbishop Lipscomb's initial reply, it looked hopeful. He asked for an estimated number of those who would be in attendance. Over the next few weeks, I collected around 100 signatures and forwarded them to him with a letter. While collecting signatures, I found two priests who volunteered to offer the Mass if it was approved. A couple of weeks later the response came. I was anxious when I opened the letter, because this process had taken quite some time and effort. I was going to relocate again in a few months, and it would be nice to have this "approved" Mass in place for at least our final weeks of living in Montgomery. As I read the letter I realized there was a disconnect; the archbishop did not see *Ecclesia Dei* as a mechanism by which Catholics could be reconciled to the Church. I'll let Archbishop Lipscomb speak for himself here:

> Thank you for excluding from participation in this effort those who, in a stable manner, might be attending what I can only regard as a schismatic chapel styled "Our Lady of Lourdes Roman Catholic Chapel." It is not, in fact, the Tridentine Mass that is offered there—though the rite ceremonies and other outward aspects of worship are very similar. The Tridentine Mass presupposes a vital union with the source of our faith and worship and that requires a priest not only validly ordained but one who possesses legitimate faculties to offer public worship.

This was patently false. Any validly ordained priest, in this case Fr. Leonard Giordina, may offer Mass in the old rite, in private, and the faithful have a right to be present. Despite the archbishop's words, the Mass attended by people in Montgomery was very much valid, since the priest was validly ordained. It may not have been "licit," but it was certainly valid. The Holy Father had asked "bishops and clergy everywhere" to be generous in allowing the old Mass, yet these people had no other real option. Where else were they to go? If there was in fact any "schism," it was created and fomented by the archbishop by not allowing these Catholics within his diocese an "authorized" traditional Mass, acting in defiance of Pope John Paul II who had asked that the traditional Mass be generously provided. He goes on:

> I would have a difficulty in approving usage of the Tridentine Mass on a weekly basis. Such an opportunity would separate those who would so use it from their parishes and the life of the Catholic community that is centered in parish relationships.

And that's exactly the point! How else were we to ensure the presence of Latin and Gregorian chant in the liturgy, according to Vatican II, and avoid the politicized environment of the parishes with their feminized liturgies? These people couldn't in good conscience attend Mass in the politicized setting he

THE CATHOLIC CENTER
Archdiocese of Mobile
400 Government St., P. O. Box 1966, Mobile, Alabama 36633, (205) 434-1585

April 3, 1995

Mr. David L. Sonnier
114 Teri Lane
Prattville, Alabama 36066

Dear Mr. Sonnier:

    Thanks very much for your letter of 8 March, 1995,
and the enclosed list of signatures of those interested in
attending a Tridentine Mass, authorized by the Church if such
sould be a feasibility in the Montgomery area. I'm certainly
well disposed toward such a move, especially if, as seems to
be the case, Father John Krozser who signed the petition, might
be prevailed upon to undertake the obligation. Over the next
several weeks I will be in the Montgomery area for Confirmations
and other commitments and consult with the clergy there looking
toward formal approval and the designation of a site.

    Thank you for excluding from participation in this
effort those who, in a stable manner, might be attending what
I can only regard as a schismatic chapel styled "Our Lady of
Lourdes Roman Catholic Chapel". It is not, in fact, the
Tridentine Mass that is offered there--though the rite ceremonies and other outward aspects of worship are very similar.
The Tridentine Mass presupposes a vital union with the source
of our faith and worship and that requires a priest not only
validly ordained but one who possesses legitimate faculties to
offer public worship. As I've indicated in the past, those
priests who presume to offer Mass at the above chapel do not
possess such faculties nor permission on my part to officiate
at public worship. Catholics who presume to share in such
liturgies give active and public support to a dimension of disobedience and division in the body Catholic that is of itself
very grave. Perhaps many do not realize this, and if the inci-
dence of the practice is as great as you seem, it might be
necessary for me to make a public statement in the Montgomery
area to correct it.

    I would have a difficulty in approving usage of the
Tridentine Mass on a weekly basis. Such an opportunity would
separate those who would so use it from their parishes and the
life of the Catholic community that is centered in parish relationships. As I've explained on more than one occasion, it
is for this reason I've not extended the celebration of the
Tridentine Mass in Mobile beyond the monthly time frame.

    As soon as I have answers to some of the questions
above with regard to availability of priests and an appropriate
place, I will be back in touch with you before publishing approval
for the Montgomery area. I deeply appreciate your perseverance
in this matter, may God bless and keep you in His love and in
ways that foster that unity in the faith for which Christ the
Lord prayed so powerfully.

                        Faithfully in Christ,

                        Most Reverend Oscar H. Lipscomb
                        Archbishop of Mobile

had created throughout his archdiocese, and he would rather have these, his most loyal sheep, cast out of the fold rather than provide for them in accordance with the Holy Father's guidance.

I began meeting with one of the priests who had volunteered to take responsibility for our soon-to-be-established Latin Mass community to discuss the details. As time dragged on, however, it became clear that the archbishop was playing a waiting game. Perhaps he discovered that I was in the military, and that I would soon be reassigned. He held out, and to ensure that we didn't go ahead without his permission he transferred the priest I had been meeting with to an obscure location on the outer reaches of the diocese, even before I finished Air Command and Staff College.

Our third child, Clair, was born in February 1995. Her birth made it easier for us to make the long commute to Birmingham for Mass. On one of these trips, we came to know Joe and Peggy Bonometti. We instantly became lifelong friends. They, like us, had been through a major spiritual transformation after having discovered that everything they had heard about Vatican II was misleading, if not outright lies. I had a lot in common with Joe, who was a 1982 West Point graduate and was currently pursuing a Ph.D. in Mechanical Engineering while working for NASA. They had been raised in an extremely liberal post-Vatican II climate, which they both found revolting but didn't talk about much since they had found peace and tranquility in becoming orthodox Catholics.

Joe confided to me on one occasion that while he was on active duty his dog tags had "Vulcan" for his religion instead of "Catholic."

"They let you do that?" I asked incredulously.

"Sure thing," he said. "An Army chaplain told me that it was his job to do whatever he could to provide for the spiritual needs of every service member. Even sci-fi 'vulcanism,' whatever that is."

This was good to know. If they could meet the spiritual needs of the "Vulcans," they should be able to meet the spiritual needs of tradition-minded Catholics.

We left in June of that year for Fort Bragg, North Carolina. The archbishop of Mobile quit responding to our requests and inquiries, and I never heard from him again.[2]

---

[2] Actually I did hear from him, years later, after I had left the Army. He wrote to me asking for donations. In my reply I told him that I would be as generous with his request as he had been with mine.

CHAPTER 8

FORT BRAGG, NORTH CAROLINA

I REPORTED TO FORT BRAGG FOR THE SECOND time in June of 1995. Fort Bragg lies next to Fayetteville, about a two-hour drive south of Raleigh. Adjacent to Fort Bragg was Pope Air Force Base (now Pope Army Airfield), which provided much of the airlift support for the XVIII Airborne Corps, the 82nd Airborne Division, and the various Special Operations units at Fort Bragg.

Despite my skepticism about any continued need for my services in Special Forces, I found that I was looking forward to this new assignment. By the time of my arrival in 1995 I had a different view of my military service to the United States. From 1984 to 1988 I had served in Special Forces during the Cold War, at a time when the mission was clearly defined. In 1988, while I was still with 7th Group, Special Forces had become a branch for officer assignments, management, and promotions. This was a significant change in that it destigmatized Special Forces, previously seen as a poor career path. Prior to this change, serving as a Green Beret officer was regarded as a career dead end—you gave up any chance of becoming a general. Now it was a good career choice, and many career-minded officers had come pouring into the Special Forces officer ranks. By a strange detour, I had stumbled back into what turned out to be an excellent career move.

However, the mission of Special Forces had become fuzzy. Special Forces operations had always fallen into four general categories, all related to the geopolitical situation that had existed during the Cold War:

- Foreign Internal Defense and Development (FIDD). This was the provision of assistance to the armed forces of nations friendly to the U.S. FIDD operations involved equipping and training combat units for combat with hostile guerrilla forces within their own countries.

- Unconventional Warfare (UW). The skills required for this type of operation were related in many ways to those required for FIDD. But in this case we provided equipment and training to an "unconventional warfare" element that was struggling to overthrow a hostile government. Operations of this nature were kept under the tightest security possible.

- Strategic Reconnaissance: A-Teams were to be capable of reconnaissance deep into enemy territory. An operation of this nature would require some means of infiltration—perhaps a waterborne infiltration or high-altitude parachute drop.

- Strike Operations: A-Teams were also to train for ambushes, raids, and other surprise assaults on a significant enemy target that other U.S. Armed Forces units were not adequately suited to assault.

These were the four classic Special Forces operations categories. Traditionally some A-Teams had been required to train more in one area than the other three, although the skills required for FIDD and UW were largely the same. In recent years the Ranger Battalions had begun to lay claim to the Strike Operations category. Now that the Cold War had ended and the Soviet Union had collapsed, the mission was not as clear. The proxy wars and revolutions that the Soviets had fostered throughout the world were a thing of the past. There was no longer the simple clarity of four general categories of operations. It appeared that now we had some doctrinal instability, and we were as likely to be tossed into such missions as "disaster assessment and survey" after a serious hurricane, "disaster relief," or election supervision and monitoring. The language abilities of the various A-Teams came in handy during conflicts like Desert Storm (Iraq) and Operation Uphold Democracy (Haiti) but clarity of mission was lacking in the new geopolitical landscape.

Despite the post-Cold War "mission creep," as this instability was referred to, it was refreshing and invigorating to be among troops again. The euphoria quickly ended with a series of sobering events. One occurred right after I reported for duty. A Special Forces soldier had committed suicide, and I was assigned the duties of "Survivor Assistance Officer," which meant helping the family to close out all of his affairs. As I prepared for the movers to pack up his belongings, I discovered items which, according to the guidelines, I was to dispose of discreetly so as to avoid embarrassing the already grieving family. There was a large walk-in closet full of hard-core pornography. It was mostly video cassettes and magazines. I nearly broke the axel of my pick-up truck carrying it all to a dump, and it took the better part of a morning, leaving me drenched with sweat, nauseated, and disgusted. I returned to the apartment to await the movers and collapsed in a chair. My eyes wandered across his desk and landed on his dog tags. His religion read, "Roman Catholic."

Was this something we could include among the "Fruits of Vatican II"? Here was another non-practicing Catholic who still identified with the Church of his childhood. What had alienated him from the practice of his faith? Had he been left feeling abandoned in the modern Catholic Church, as I had? Had the public face of the Catholic Church become so feminized and weird that a young man who was willing to lay down his life for his country would have nothing to do with it? Had the faith that should have sustained him instead been taken from him? Did he feel alienated as a warrior who was also Catholic? Before my awakening I would not have even thought of these things. Now my heart ached for this young man, near my age, who had been raised a Catholic as I had. He had joined the Special Forces, was willing to risk his life in the defense of his country, and had somehow reached a level of despair that brought him to suicide. As I sat there, I wished that I could have been available to get to know him, talk to him, help him to understand what had happened with the Church and maybe give him some books to read or help him to learn some prayers in Latin. I wished I could have been there to help him reject the temptation to despair and instead rediscover the faith of his fathers. This was a sad return to Fort Bragg.

I was assigned as the commander of B Company, 2nd Battalion, 3rd Special Forces Group. Our battalion was in Haiti at the time I signed in, and I would end up making several trips there over the coming months. Part of the battalion was left behind at Fort Bragg, and we carried out administrative duties that supported the operation of the deployed element. Our battalion was working to restore order, get water supplies running again, supervise elections, monitor construction projects, and work with Civil Affairs units. These units, trained to provide and rebuild social infrastructure, were instead passing out coloring books to teach young Haitians about birth control. This is what happens in an Army full of Catholics who don't know their faith.

That Sunday we went to Mass in one of the Fort Bragg chapels, only to find that the agenda was firmly entrenched. There were girl altar boys, the sermon was a political message that leaned feminist, and the Scripture readings were once again being altered. The music was ugly, horrible! Felt banners were flying. No wonder we had soldiers passing out birth control coloring books in Haiti and committing suicide, and covering their bodies with grotesque tattoos. I was noticing things that would have escaped my observation had I not discovered that the Church in the U.S. had been hijacked. Ever the servant of God and the United States Army, however, I did not despair. I discussed it all extensively with Lorri and we saw it as my duty to try to elevate the culture a bit; to encourage Catholic soldiers to be good Catholics; and to try to help my soldiers discover their moral compass. I also began looking into the issue of Church authority at this point. The Fort Bragg chapel activities that come under the title of "Catholic" are under the jurisdiction of the Archdiocese for the Military Services, which at that time was led by Archbishop Joseph Dimino. Perhaps I would write to him to request that the traditional Latin Mass be available for those servicemen who felt out of place in the modern Catholic world. This could only be a positive influence in the lives of these poor soldiers. Meanwhile, Lorri and I would look into the local Maronite rite parish.

Among the things I had learned in the past couple of years was the fact that there are many rites in the Catholic Church, and most of them had not gone through the insane convulsions that affected the "Latin rite" in which Latin was largely forbidden. The Eastern rites are Catholic, in every sense of the word, although their liturgy is different from that of the Roman rite. They often celebrate different feast days, they have distinct devotions, practices, and vestments, and they even have a separate code of canon law. In Fayetteville, Archangel Michael Maronite Catholic Church did not fall under the jurisdiction of the bishop of Raleigh, but rather a Maronite Catholic bishop in New York. Above him in the hierarchy is a patriarch of the Maronite Catholic Church, who resides in Lebanon. Above him, of course, is the pope, which is why the Maronite Church is

"Catholic." As I discovered, many Roman Catholics sought relief in Oriental rites of the Church after the spread of modernism throughout the Roman rite. The Eastern Catholics are often rather insular because they exist for an immigrant group. In this case, several Lebanese families were at the center of the parish life. Communities of Eastern Catholics are usually not as influenced by modern culture, so their liturgy, spirituality, and prayer life tend to remain much closer to the heart of the Church. They were conspicuously lacking in political agendas. And since their liturgies did not fall under the Roman rite, they were not subject to the whims of a liberal liturgist or a modernist bishop. Hoping to avoid the Novus Ordo problems we had left behind in Alabama, we decided to make this Maronite parish our own for the time being—at least until we could somehow secure permission for a Latin Mass on base or somewhere nearby.

Coincidentally, Pope John Paul II had just published a document entitled *Orientale Lumen*, or "Light of the East," in which he addressed the role of the Eastern rites in the Catholic Church.[1] The Maronites have a beautiful chanted liturgy. Their Eucharistic prayers are in Syriac, which is like Aramaic, the language Christ himself spoke at the last supper. We found this branch of Catholicism to be quite attractive. Uniquely among the Eastern Catholic Churches, the Maronite Church is known by the name of a person—St. Maron, a fourth century hermit. By his holiness and miracles he attracted many followers. When he died in AD 410, his disciples built a large monastery in his honor, from which other monasteries were founded. St. Maron's followers were always faithful to the Catholic Church, and they maintained their bonds with Rome and every Successor of St. Peter throughout their history. During the seventh century, the Maronites suffered persecution and sought refuge in the mountains of Lebanon. There they continued to maintain their liturgy, faith, culture, and spirituality largely unchanged.

We instantly fell in love with the Maronite parish, but we are not Maronite Catholics. We are Roman rite Catholics whose "rightful aspirations," to use the words of the pope himself, were to worship in the way our parents and ancestors had always worshipped. Some will accuse Catholics who attend a parish of a

---

[1] Apostolic Letter *Orientale Lumen*, May 2, 1995, marking the centenary of Leo XIII's Apostolic Letter *Orientalium Dignitas*.

different rite of "rite shopping." These people will claim that we were wrong to make ourselves at home among the Maronites. But the first law of the Church is the salvation of souls, and we found the highly politicized Novus Ordo distressing to the point of being a scandal. It was potentially a stumbling block for our own salvation. As much as we loved the Maronite rite, the traditional Latin Mass had not been re-established in the Diocese of Raleigh as Pope John Paul II had requested, and this presented a problem for us. I decided to send a polite letter to Bishop Gossman of the Diocese of Raleigh requesting that the traditional Latin Mass be made available in his diocese.

The bishop of Raleigh wrote back promptly. He simply instructed me to take the issue up with the Military Archdiocese, since I was a member of the U.S. Army. I didn't think it appropriate to respond asking what he was doing to respect the Holy Father's will, as expressed in *Ecclesia Dei*. At least not yet.

I sent a letter to the archbishop of the Archdiocese for Military Services and put forth the same request, asking that the traditional Latin Mass be made available at one of the many chapels on the huge installation. A couple of weeks later the response arrived, signed by the Vicar General of the Military Archdiocese:

> Before a judgment can be made it is necessary to hear from the chaplain(s) at the bases about this matter. The request must come from the senior Catholic chaplain who in the meantime should contact this office for the information that would be needed. I would therefore suggest that you talk this over with the Catholic chaplains of the two bases, if you have not already done so.

Fair enough. One of the chapels at Fort Bragg was specifically a Catholic chapel. I did some research, and what I found was that there were six Catholic chaplains assigned to Fort Bragg and one additional at the adjoining Pope Air Force Base. That may seem like a lot, but for a military community of 42,000, some thirty percent of whom identified as Catholic, that adds up to only one priest for every 2,100 Catholic soldiers and airmen, not including their families. The ratio of priests to laymen had declined rapidly throughout the military since Vatican II. Soon the number of Catholic chaplains on active duty would fall below 100, down nearly a third of the 286 needed by the active Army. By now it was well known that there was a serious shortage of

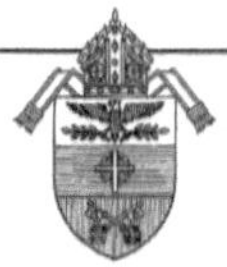

ARCHDIOCESE FOR THE MILITARY SERVICES, USA

Telephone: (301) 853-0400
Fax: (301) 853-2246

Post Office Box 4469
Washington, D.C. 20017-0469

Major David L. Sonnier, USA
P.O. Box 74609
Fort Bragg, NC 28307-4609

Dear Major Sonnier:

Archbishop Dimino received your letter of 3 July regarding the possibility of a Tridentine Mass at Fort Bragg and Pope Air Force Base.

Before a judgment can be made it is necessary to hear from the chaplain(s) at the bases about this matter. The request must come from the senior Catholic chaplain who in the meantime should contact this office for the information that would be needed.

I would therefore suggest that you talk this over with the Catholic chaplains of the two bases, if you have not already done so.

With every best wish, I am

Sincerely yours in Christ,

(Rev.) Nicholas Halligan, O.P.
Vicar General

August 1, 1995

Serving Catholics in the Army, Navy, Air Force, Marine Corps, Coast Guard,
V.A. Medical Centers and Government Service Overseas

priests everywhere, and the military was to be more immediately affected by this trend, because it's a young man's profession. The idea that it might have something to do with the disastrous state of the liturgy and doctrine since Vatican II seemed to have occurred to absolutely nobody.

I met with the senior Catholic chaplain, a graying older priest from New York who held the rank of Colonel and seemed to have painted his hair black to avoid the appearance of aging. He was preparing for retirement soon. He seemed uninterested in my request for a Latin Mass, sidestepping the question by reminiscing about his stellar (preconciliar) Catholic education. I could see no point in his telling me this. Sure, he had a great education. This was what I wanted for my children, and yet it was virtually impossible to find a good Catholic education now. That's why Catholic parents were choosing to homeschool, Lorri and I included. I kept pressing the question.

"Things are better now than before Vatican II," he explained.

"What about all the shortages of priests, empty seminaries, and declining orders of nuns?," I asked.

"Oh, well, that." He waved. "Well, as a pastor, I have to say that things are much better."

I pressed on, trying to coax a response from him. Perhaps he would be interested in offering the Mass for us himself until he retired. He suddenly became very agitated and anxious to get rid of me. I could tell that I wouldn't get anywhere with him, so I decided that it would be best to pray and wait for his replacement to arrive.

The wait turned out to be several more months. When his replacement finally did arrive, I made an appointment to visit our new senior Catholic chaplain, Fr. (Col.) Sidney Marceaux. Fr. Marceaux had no knowledge of the purpose of our meeting. I was met with silence when I inquired about the possibility of having a traditional Mass celebrated somewhere on the installation on Sundays and Holy Days. I sensed hostility immediately.

"For what purpose?"

"Because the Holy Father has asked bishops and all pastors to be generous in allowing the old liturgy of the Roman rite, and it's beautiful. So, why not?"

"What's wrong with the new liturgy? Isn't it beautiful too?"

How could I respond to this? Fr. Frederick Faber, a priest at the Brompton Oratory in London who lived from 1814–1863, had famously called the Mass "the most beautiful thing this side of heaven"—but he was referring to the traditional Latin Mass which was nowhere to be found these days. I was tempted to go through a litany of the problems of the new Mass, but instead I put that thought away and asked God to help me to do my best to answer without lying.

"Well … uh … I guess, but we've come to appreciate the Latin Mass."

"You're too young to remember the Latin Mass."

"No, I'm not too young. I could remember the one in Dayton, Ohio, and the one in Birmingham, Alabama. I have attended them within the past two or three years. The local bishops in those locations gave permission for the faithful to follow the old rite. The bishop in Mississippi who confirmed me offers a Latin Mass in his Cathedral one Sunday a month.[2] They do this because the pope asked them to. He's asking you, too."

"No, that's only for old people who grew up with it. When they die, the old Latin Mass will die with them. You didn't grow up with the Latin Mass; it's not intended for you. We have a new Mass now, and that's the one we're to follow."

At this point I thought that honesty would be the best approach.

"Well, Father, I did that for a while, but there seems to be no end to the tinkering with it by the liberals. They don't know when to stop. Anyway, populating the altar with women is a step too far. I don't want my children seeing that."

In retrospect, honesty was not the best approach. This seemed to anger him greatly, but he tried not to show it. He demanded that I put the request in writing, which I did immediately. Along with the request, I sent a copy of the pope's document *Ecclesia Dei*, information about the *Fraternitis Sacerdotalis Sancti Petri* (FSSP, a society of priests established by John Paul II for those desiring to follow the traditional Latin rite), and a list of "authorized" Latin Masses.

His response was short: "After consulting with the Catholic chaplains assigned to Fort Bragg, I have decided not to grant your request."

Of course, he gave no reason. I had no intention of accepting this response, because the Holy Father's plea for generosity could in no way be considered fulfilled when there was not a single authorized Latin Mass anywhere in the state of North Carolina.

---

[2] Bishop Joseph L. Howze, Diocese of Biloxi. Bishop Howze gave the first-ever "universal indult" in his diocese. He told his priests that any of them could offer Mass in the old rite any Sunday of the month in their own parishes except for the first Sunday—and that was because he wanted to have a gathering once a month in which all the followers of the traditional rite could come to the cathedral for a High Mass! None of the priests took him up on it, so he offered a Low Mass once a month in a side chapel of the cathedral until he retired.

Fortunately, since we had the Maronite Church nearby, there was no sense of extreme urgency to get permission pushed through; we were able to attend Divine Liturgy there in tranquility, thereby fulfilling our Sunday obligation without suffering ideological assaults on our sensibilities. My wife and children attended the daily Novus Ordo Masses in a Fort Bragg chapel offered by a retired "contract" priest who, despite following the newer liturgy, seemed to have a prudent sensibility. The dearth of military chaplains had led to hiring local or retired priests.

As soon as I had the opportunity, I made an appointment with Fr. (Col.) Marceaux to see what his objection was, since it was not stated in his response. He was late, so I waited. And waited. And waited.

Now, let's put this in perspective. I was a Major in the U.S. Army Special Forces, and a commander of Company B, Second Battalion, 3rd Special Forces Group (Airborne). My time was a most precious asset; I had so little of it that it was difficult for me to be sitting in this office to begin with. I had abandoned my company for at least an hour on a very busy day. I have no recollection of what my company was doing on that day, but on any given day we had A-Teams on one of the ranges firing mortars or weapons of some make and caliber. Perhaps we were in the field somewhere, studying Spanish or French, doing some kind of training having to do with the numerous perishable Special Forces skills, or doing some form of maintenance on our expensive equipment. Yet, for the sake of the faith of my children, my wife, myself and the Catholic soldiers under my care, I sat here in hopes of bringing something truly beautiful to this Army I loved. An opportunity that could give these folks a better alternative for attending Mass on Sundays and Holy Days. Something to help them find greater sanctity and strength of soul as they prepared for the dangerous tasks they might be called to undertake at a moment's notice.

Fort Bragg is the heart of the U.S. Army. The 82nd Airborne Division is there, as well as its higher headquarters, the 18th Airborne Corps; two Special Forces groups were there, the 3rd and the 7th, as well as our higher headquarters, the U.S. Army Special Operations Command. Many soldiers were Catholic. It was abundantly clear to me by now that many of them had abandoned the practices of their faith. It is not difficult to understand the reason for this. You don't tell Green Berets and Paratroopers they have to hold hands and sing folk songs at Mass. You don't feminize the liturgy in such a testosterone-laden setting and expect them to accept it. Protestant churches in the area were full of fallen-away Catholics. Northwood Temple claimed more than four hundred Catholic families in their congregation.

The bishop of Raleigh occasionally called our Maronite priest to tell him that only Maronite Catholics were permitted to attend Mass at Archangel Michael Maronite Catholic Church. He demanded that the priest stop Latin-rite Catholics from attending Mass there. This was a most ridiculous attempt at abuse of ecclesiastical authority, as he had no right to mandate such a thing. Any Catholic can attend Mass in the Eastern rites — and should consider doing so at times like this. Finally, during one such phone call our Maronite priest told him, "Why don't you contact Northwood Temple and tell them what you're telling me? There are far more of your Roman-rite Catholics over there!"

Finally, Fr. Marceaux arrived. Making no apology for his tardiness, he called me into his office with a flip of the wrist. I stuffed my green beret in my right-side pocket, sat and waited for him to get settled; I listened to his complaints about his back problems and waited patiently for him to get to the point.

"What brings you here?"

I asked him about his response; why was he not giving us permission to have a Latin Mass?

"Well, it would not be a good idea."

"Why not?"

"Your thinking is not in line with the USCCB [United States Conference of Catholic Bishops]. Therefore, I can't grant your request."

"Why do you say that? My thinking is 'not in line with the USCCB?' What does that mean?"

"Because you're opposed to having women on the altar."

The words hit me rather hard. Later on, I would find it comforting that the accusation was that my thinking was "not in line with the USCCB" instead of "not in line with the pope" or "not in line with the Universal Church," but at the time it felt like this priest was accusing me of being a dissident, or a heretic. I brushed it off and continued, because there was much more at stake here than my own personal feelings.

"Fair enough, but what good results do we see coming forth from this kind of liturgical experimentation? Where's the 'new springtime' we keep hearing about? Declining vocations everywhere. And these girls will never eventually be ordained, so it's best left to boys and men to serve at the altar."

"That's not necessarily true. Some canon lawyers believe that women can become deacons. In fact, some of the best canon lawyers in the United States are convinced it will happen soon."

Fr. Marceaux was a canon lawyer. At least he called himself one. I decided not to argue with him about what the "best canon lawyers in the United States" believe.

"What about having a Latin Rosary?" I pursued. "What about having the New Mass in Latin, or at least partially in Latin with Gregorian chant, the way that the Second Vatican Council stated in the document on the liturgy?"

"No. That might lead to other things."

"What other things?"

Silence.

"What other things?" I asked again.

"It would be divisive."

"How can something the Holy Father asked you to do for the faithful who request it be divisive?" I asked.

"I'd rather not talk about it."

"Why not?"

"I'd rather not say."

It was clear that this priest was not impressed with the words of Pope John Paul II, and it was clear that he was not going to allow me to have any input on Catholic activities taking place on Fort Bragg. Even though he was ignoring the Holy Father's guidance, he somehow considered me to be the dissident, since my thinking was "not in line" with the U.S. Catholic bishops.

In my heart I knew that I was not a dissident, and that my loyalty was to the Holy Catholic Church of Rome, whose present Vicar, Pope John Paul II, was having a most difficult time with dissidents who really were in rebellion—perhaps people like Fr. Marceaux! Whatever happened with the Church in the U.S., my loyalties were to remain with the Church of Rome. Therefore, I decided simply to appeal through ecclesiastical channels.

First, to ensure that I had tried all avenues at the local level, I contacted the Catholic Chaplain at Pope Air Force Base and provided him with some information about *Ecclesia Dei* and a request for a weekly Mass according to the 1962 Missal. The letter included a request for the new Mass in Latin, should the use of the 1962 Missal be impossible. The result was predictable:

9 October 1996

SUBJECT: Latin Mass

1. After our previous phone conversations and your mailing me your package asking about the possibility of having a Latin Mass here at Pope I am responding to your request. After prayerful consideration I have decided not to grant your request.

2. I have strong convictions about fragmenting the Catholic community. The Mass is a sign of our unity and to begin to separate that unity because of special needs creates an atmosphere of individualism and special interest that weakens the bond of unity and can become divisive. I also have a strong working relationship with the Catholic priests at Fort Bragg and since you are a member of that community, I do not wish to undermine their authority in the decision that they reached. To date, no member of the Catholic community at Pope AFB has requested to have a Latin Mass available.

3. It is unfortunate that we have lost touch with the Latin liturgical tradition. I believe a better way of tapping into this liturgical tradition is to integrate more traditional Latin music into our celebration of Mass. The Church should use her whole treasury of music when celebrating Eucharist. I personally miss some of the wonderful Gregorian chants in Latin that were used in the seminary. In my previous parish assignments I worked closely with music ministers and encouraged them to use more traditional Latin music. I found the music of Taize to be both beautiful and fairly easy to have a choir sing.

4. By copy of this memorandum, I am informing the Archbishop of the Archdiocese for the Military Services, of my decision and am sending a copy to your pastor, Fr. Marceaux.

Thomas M. Angelo, Ch, Capt, USAF
Senior Catholic Chaplain

DEPARTMENT OF THE AIR FORCE
HEADQUARTERS 23D WING (ACC)
POPE AIR FORCE BASE, NORTH CAROLINA

9 October 1996

MEMORANDUM FOR Maj, David L. Sonnier, 2d Battalion, 3d Special Forces Group, Fort Bragg, NC 28307

FROM: 23 WING/HC
315 ETHRIDGE ST.
POPE AFB NC 28308-2396

SUBJECT: Latin Mass

1. After our previous phones conversations and your mailing me your package asking about the possibility of having a Latin Mass here at Pope I am responding to your request. After prayerful consideration I have decided not to grant your request.

2. I have strong convictions about fragmenting the Catholic community. The Mass is a sign of our unity and to begin to separate that unity because of special needs creates an atmosphere of individualism and special interest that weakens the bond of unity and can become divisive. I also have a strong working relationship with the Catholic priests at Fort Bragg and since you are a member of that community, I do not wish to undermine their authority in the decision that they reached. To date, no member of the Catholic community at Pope AFB has requested to have a Latin Mass available.

3. It is unfortunate that we have lost touch with the Latin liturgical tradition. I believe a better way of tapping into this liturgical tradition is to integrate more traditional Latin music in our celebration of Mass. The Church should use her whole treasury of music when celebrating Eucharist. I personally miss some of the wonderful Gregorian chants in Latin that were used in the seminary. In my previous parish assignments I worked closely with music ministers and encouraged them to use more traditional Latin music. I found the music of Taize to be both beautiful and fairly easy to have a choir sing.

4. By copy of this memorandum, I am informing the Archbishop of the Archdiocese for the Military Services, USA of my decision and am sending a copy to your pastor, Fr. Marceaux.

THOMAS M. ANGELO, Ch, Capt, USAF
Senior Catholic Chaplain

CF:
Most Reverend Joseph T. Dimino
Ch, Col, Sidney Marceaux

Global Power for America

Several things are obvious from this letter. There had been no concern about damaging the "unity" of the Catholic Mass by promoting the "special needs" of feminists. In the third paragraph, the chaplain pointed out, correctly, that Latin belongs in the liturgy, but it is quite impossible to sing Gregorian chant if attending Mass to do so means accepting the presence of girls at the altar, or strident, angry women reading modified feminist-friendly Scripture.

At the same time, a multitude of liturgies were being allowed: Children's Masses, guitar Masses, and Spanish Masses. How could a Latin Mass further "fragment the Catholic community?"

Basically, this priest was saying, "it's sad that the beauty of the liturgy is lost, but it's lost and, for the sake of 'unity' we all have to wander in the wilderness together singing Kumbaya."

"Go ahead," I thought, "but don't expect me to follow." In March of 1996 I sent an appeal to the office of Archbishop Dimino, at the Archdiocese for Military Services:

> I mailed a letter to you in July 1995, in which I requested that the Tridentine Mass be made available to Catholics at Fort Bragg and Pope Air Force Base. A copy of this letter is enclosed, as well as the response from your office indicating that we should coordinate this effort through the chaplains at Fort Bragg and Pope AFB. We have made every effort to do so, but the senior chaplains are opposed to the idea. I anticipate that you will have no request for permission coming forth from them. This does not mean that the need is not here, it simply means that the senior Catholic chaplains do not support our request.
>
> Although they do not understand our concerns, our Holy Father, Pope John Paul II has indicated that *he* does, and he has asked for the wide and generous support from bishops and all others engaged in the pastoral ministry in the Church.

The reply arrived later that month:

> Dear David,
>
> I wish to acknowledge your recent letter to Archbishop Dimino regarding your request for the "Tridentine Mass." As you have already been informed by Father Halligan in his letter to you of August 1, 1995, any such request should be presented by the Catholic chaplain for consideration.
>
> With every best wish, I am
>
> Sincerely Yours in Christ,
> Rev. Msgr. Aloysius R. Callaghan, S.T.L., J.C.D.
> Moderator of the Curia/Chancellor

What was *this*? Did he even bother to read the letter I sent him? The Catholic chaplain had denied our request, and I was making an appeal to him to assist. Perhaps to make a decision that would overrule the decision of Fr. Marceaux. The Holy Father had asked for generosity and there wasn't an authorized Latin Mass anywhere in the state of North Carolina. When the Holy Father asks for the support of the "bishops and all those engaged in the pastoral ministry in the Church" on something, you would think that it would be respected.

I couldn't let this response stand. There was obviously a disconnect here.

Meanwhile, my wife had become an active member of the Military Council of Catholic Women (MCCW) and she met a retired priest that lived in the area. Msgr. Edward Spiers was a kind, older priest, certainly old enough to remember the Church's better days. We hoped he would be willing to support our Latin Mass community once we organized it. When I contacted him, not only was he willing, he was enthusiastic about the idea!

Soon I began to meet other Catholics who would be interested in being a part of a Latin Mass community. In some cases they had attended the Latin Mass in other locations around the country, and they were just as frustrated with the situation as we were.

I contacted Fr. Marceaux again. Perhaps if he knew that a retired priest was offering to help bring the Latin Mass to Fort Bragg, he would be more willing to approve it. Maybe he didn't want to burden the already overworked chaplains, but with a retired priest taking responsibility, surely, he would allow us to proceed.

The answer was definitely "no," and he seemed quite upset by my persistence. And "no," he would not allow Msgr. Spiers any such role among the Catholics at Fort Bragg either.

Now, at this point, we must wonder "why not." What harm could possibly have come from a retired old priest coming to Fort Bragg and offering Mass in one of the multitude of chapels at the garrison on Sundays and Holy Days? There were other retired priests who played an active role; the daily Mass and Saturday vigil Mass at our own USASOC chapel was offered by a retired priest. What could possibly be the harm? He wouldn't say.

Soon after this we heard through some of the women in the MCCW that Fr. Marceaux and other priests were warning Catholics to "beware of people trying to cause divisions" in the Catholic community at Fort Bragg. They were now issuing warnings from the pulpit that there were people who "couldn't cope with change" trying to throw the Church back into an earlier era and trying to divide the Church. These dangerous dissidents were circulating a petition that Catholics should refuse to sign.

We got a laugh out of it, but it seemed peculiar. With all the high-profile Catholic pro-abortion politicians, the *real* public dissidents who opposed the Church on numerous fundamental issues, why would they target *us* for calumny—faithful Catholics who were only trying to exercise a valid option for worship? All we were asking for was for an option, a valid option. And unlike the girl altar boy option which had been a strictly reserved permission, this was an option for which the Holy Father himself had implored generosity in a motu proprio.

I wrote to the Archdiocese for Military Services again in May, 1996. Perhaps if the archbishop knew that we had a priest to volunteer for the effort, he would be more willing to approve our request, override the senior chaplain's decision, or take some meaningful action on our behalf. Before sending the letter, I showed it to Msgr. Spiers and made a few editorial changes he suggested:

24 May 1996

Archbishop Dimino:

I have appealed to your office on several occasions, on behalf of myself and others stationed at Fort Bragg, to request that the traditional Latin Mass be made available here under the provisions of *Ecclesia Dei*. You relinquished the decision for implementation of *Ecclesia Dei* to a local level, specifically, to the Garrison Chaplain, who has repeatedly denied our requests without giving any good reason. The only reason offered by the Senior Catholic Chaplain is that it would be 'divisive,' as if compliance with something the Pope has asked 'wide and generous' support for can possibly be divisive. Fort Bragg, as you know, has one of the largest concentrations of troops anywhere, and so it is no surprise that a significant number of Roman Catholics here who grew up with the Latin Liturgy desire to have the traditional Latin Mass.

Now there is a retired priest, Msgr. Edward Spiers, volunteering to have the traditional Latin Mass here at Fort Bragg weekly, and yet the only thing stopping him is that Chaplain (Col.) Marceaux, the Garrison Chaplain, is not giving us permission.

I think that I am justified in appealing his decision. I am appealing to you, with hopes that you can assure

Chaplain (Col.) Marceaux that the traditional Latin Mass is not bad or divisive in any way, but rather it is something that will bring Roman Catholics together to a greater understanding of the Church and our traditions.

I would suggest, rather than a *reversal* of Chaplain (Col.) Marceaux's decision, a *modification*, in which the traditional Liturgy is allowed for a period of several months, during which the impact is evaluated to determine whether it is good or bad.

Certainly, the problem should be clear to the archbishop now. Giving him the benefit of the doubt, as Msgr. Spiers said I should do, I prepared for a long, patient wait for him to respond. Once he understood the problem, he would certainly want to spend some time thinking about it.

It took only four days for the response to arrive:

28 May 1996

Dear David,

I wish to acknowledge your letter of May 24, 1996 written to Archbishop Dimino.

As for the request for the traditional Latin Mass, I once again wish to clarify that such a request must be presented by the Catholic Chaplain.

With every best wish, I am

Yours in Christ,

Rev. Msgr. Aloysius R. Callaghan, S.T.L., J.C.D.
Moderator of the Curia/Chancellor

This bureaucratic runaround was becoming rather difficult to believe. The Church truly must be a divine institution to survive such incompetence! Had he not even read what I sent to him? I was trying to appeal to the archbishop, and he was referring me back to the same chaplain whose decision I was appealing! Furthermore, the response came from this "Moderator of the Curia/Chancellor" instead of the archbishop. Again.

I gave a copy of the response to Msgr. Spiers. He was surprised by it, but being more than eighty years old, there wasn't much in the world that could upset him. "Don't worry about it," he said. "I'll take care of getting permission for the Mass."

May God bless Msgr. Spiers for his kindness and his willingness to assist us. His letter is reproduced here:

ST JOSEPH OF THE PINES, INC.

Most Reverend Joseph T. Dimino, D.D.   570 Central Drive
Archbishop of the Military Services   Southern Pines, NC 28387-2899
962 Wayne Ave.   (910) 692-2212
Silver Springs, MD 20910   *A member of the Sisters
                                                 of Providence Health System*

Most Reverend and dear Archbishop   10 July 1996

INTRODUCTION: I am a retired priest (Columbus,OH diocese),
ordained in '39, now living at St. Joseph Hospital and
Retirement Villas with part-time chaplain duties and also
serving in my sixth year as contract chaplain at Ft. Bragg
which is adjacent to the hospital wheree I reside. I have
been frequently requested to assist with the Sunday Mass
schedule at Bragg where 13 services are scheduled/Sunday.
When the IRAQ War developed I received a routine call to
assist the following Sunday; it was however anything but
routine. I arrived at the designated chapel about 30   + Confession.
minutes early where I was scheduled for two Masses at the
Main Post Chapel. I found a huge crowd surrounding the
chapel. When I arrived several senior officers complained
about the absence of any chaplains from all the scheduled
Masses. After assuring them I did not know the reason, I
invited them all to attend my Mass at which I left the
doors and windows open since the crowd was too large for
the chapel. I also agreed to offer several additional
Masses and had this info and times posted in the chapels
where Masses were scheduled.

                        pr Bragg
As you know, five(military chaplains (all priests) were
deployed to Saudi Arabia in the first wave. The info for
security could not have been given to me. Though retired
and at that time of 75 years in age, I recognized the need
and opportunity to serve God and country And have continued
as a contract chaplain though the remuneration barely
covered car expenses. In addition, I recruited other diocesan priests to cover the 13 Sunday Masses. Following the
war's end, I was requested and have continued to allow the
returning military chaplains the opportunity to using their
accumulated vacation leaves. At the present time due to
a change in the Mass schedule, I am not needed but Fr.
Sidney Marceaux has requested and I have agreed to continued
service.

(continued)

10 July 1996

Most Reverend and dear Archbishop:

I am a retired priest (Columbus, OH diocese), ordained
in '39, now living at St. Joseph Hospital and Retirement
Villas with part-time chaplain duties and also serving
in my sixth year as contract chaplain at Ft. Bragg which
is adjacent to the hospital where I reside. I have been
frequently requested to assist with the Sunday Mass
schedule at Bragg where 13 services are scheduled/Sun-
day. When the Iraq War developed, I received a routine

call to assist the following Sunday; it was however anything but routine. I arrived at the designated chapel about 30 minutes early where I was scheduled for two Masses and Confession at the Main Post Chapel. I found a huge crowd surrounding the chapel. When I arrived, several senior officers complained about the absence of any chaplains from all the scheduled Masses. After assuring them I did not know the reason, I invited them all to attend my Mass. I also agreed to offer several additional Masses and had this info and times posted in the chapels where Masses were scheduled.

As you know, five Ft. Bragg military chaplains (all priests) were deployed to Saudi Arabia in the first wave. The info for security could not have been given to me. Though retired and at that time of 75 years in age, I recognized the need and opportunity to serve God and country and have continued as a contract chaplain though the remuneration barely covered car expenses. In addition, I recruited other diocesan priests to cover the 13 Sunday Masses. Following the war's end, I was requested and have continued to serve to allow the returning military chaplains the opportunity to use their accumulated vacation leaves. At the present time due to a change in the Mass schedule, I am not needed but Fr. Sidney Marceaux has requested and I have agreed to continue service.

Recently I have been interviewed by several senior officers and their families if I would be available to offer the Mass which Pope John Paul II in his Apostolic letter *Ecclesia Dei* has urged Bishops worldwide to grant permission.

I answered by explaining: a) such permission for military installations must be obtained from Your Eminence as the Ordinary for military services; b) granted that your permission is given, it would then be the responsibility of Fr. Sidney Marceaux to designate the celebrant as well as the time, chapel and other circumstances. c) Given the affirmative for both a and b, I would be willing to offer the traditional Latin Mass which I was privileged to offer the first 37 years of my priesthood and which I have greatly missed during the succeeding 20 years.

ST JOSEPH OF THE PINES, INC.

590 Central Drive
Southern Pines, NC 28387-2899
(910) 692-2212
*A member of the Sisters
of Providence Health System*

page 2: Archbishop Joseph Dimino

TRADITIONAL LATIN MASS REQUEST

Recently I have been interviewed by several senior officers and their families if I would be available to offer the Mass which Pope John Paul II in his Apostolic letter "Ecclesia Dei" has urged Bishops worldwide to grant permission.

I answered by explaining: a) such permission for military installations must be obtained from Your Eminence as the Ordinary for military services; b) granted that your permission if given, it would then be the responsibility of Fr. Sidney Marceaux to designate the celebrant as well as the time, chapel and other circumstances. c) Given the afffirmative for both a and b, I would be willing to offer the traditional Latin Mass which I was privileged to offer the first 37 years of my pristhood and which I have greatly missed during the succeeding 20 years.

As a result of my several meetings with the Army personnel who are making their request for what the current successor of St. Peter has authorized, I am highly impressed with their religious fervor, their respectful loyalty, and their knowledge of Catholic doctrine and history, I do hope that your Eminence may reply favorably to their request, perhaps limited to a probationary period.

Respectfully in Christ,

edward f spiers

rev. Msgr. edward f. spiers, Ph.D.

P.S. please pardon the typing errors; I do not rate secretarial services

efs.

As a result of my several meetings with the Army personnel who are making their request for what the current successor of St. Peter has authorized, I am highly impressed with their religious fervor, their respectful loyalty, and their knowledge of Catholic doctrine and history. I do hope that your Eminence may reply favorably to their request, perhaps limited to a probationary period.

Respectfully in Christ,

Rev. Msgr. Edward F. Spiers, Ph.D.

I left it to the good Monsignor to sort out while I focused on doing my job. A lot was going on; our battalion had returned from Haiti, and we were constantly on the move. I remained the commander of B Company, 2nd Battalion, 3rd Special Forces Group, but soon I would be turning over the company to another officer and assuming responsibilities as the battalion executive officer. In the Special Forces Battalion there are five positions for majors: command of the three "line" companies (A Company, B Company, and C Company) as well as the Operations Officer (S-3), and the Executive Officer (XO). The XO is responsible for oversight of the staff and ensuring that the battalion commander doesn't get bogged down in the minutiae required to run a battalion.

I often invited fellow Catholics to attend Divine Liturgy at the Maronite Church with us. One Sunday, I met another serviceman at the Maronite Church who turned out to be a military lawyer, a Judge Advocate General (JAG) officer. Tim (not his real name) was a major. He had been a member of an "Indult" parish in Richmond Virginia during a previous assignment. When I told him my saga with obtaining permission for an Indult Mass at Fort Bragg he became interested. "I'll write the senior chaplain a letter," he said.

Tim's letter was a masterpiece! We made copies of it and mailed them to every Catholic chaplain at Fort Bragg. It is reproduced below:

> Dear Fr. Marceaux,
>
> I am a Catholic Judge Advocate recently assigned to Fort Bragg. As a former member of Saint Joseph's Chapel (Ministry of Saint Benedict Parish) in the Diocese of Richmond, Virginia, I have a deep affection for the traditional Latin Mass. In accordance with the Holy Father's 1988 motu proprio *Ecclesia Dei*, I am writing respectfully to seek your support for a petition to the Archdiocese for Military Services requesting authorization of the traditional Latin Mass on Fort Bragg.
>
> As you know, the Holy Father decreed in *Ecclesia Dei* that "respect must everywhere be shown for the feelings of those who are attached to the Latin liturgical tradition, by a wide and generous application of the directives already issued by the Apostolic See, for the use of the Roman Missal of 1962." Those directives, issued in 1984, authorized local bishops to permit the traditional Latin Mass within their respective dioceses. It is my

understanding that His Excellency, Archbishop Dimino, defers to the wishes of post chaplains in granting such permission within the Military Archdiocese.

Opponents of the traditional Latin Mass have argued that *Ecclesia Dei* should be construed very narrowly and that permission should be granted only rarely. I believe this interpretation is inconsistent with the plain language of the Apostolic Letter ("wide and generous application"). Even so, the indult's express purpose—"to facilitate [the] ecclesiastical communion" of "those Catholic faithful who feel attached to some previous liturgical and disciplinary forms of the Latin tradition"—would be fulfilled at Fort Bragg, whence a number of soldiers travel to Raleigh and other cities to attend the Latin Mass in schismatic churches.

Many U.S. bishops, ranging across ideological lines, have liberally approved indult Masses within their dioceses. The Coalition in Support of *Ecclesia Dei*, reports that weekly indult Masses can be found at 40% of U.S. Dioceses (*The Latin Mass*, Winter 1996, at 10). These bishops include such progressives as Bishop Frank Rodimer in northern New Jersey, Cardinal Joseph Bernardin, Bishop Walter Sullivan in Richmond, Virginia (whom I have seen celebrate the traditional Latin Mass himself on occasion), and Archbishop Rembert Weakland of Milwaukee (*The Latin Mass*, Summer 1995, at 4; Winter 1996, at 58–59). Within the Military Archdiocese, I know that the traditional Latin Mass is permitted at Fort Carson, Colorado.

I believe that the indult Mass—far from being divisive—positively contributes to Church unity and vitality. It provides room within the contemporary Church for those who prefer a more structured, aesthetic, and contemplative approach to liturgy. Numerous Catholics alienated by contemporary developments in the liturgy have returned to the Church in the wake of *Ecclesia Dei*. In the Diocese of Wichita alone, for example, over 120 lapsed Catholics were brought back to the church by the indult Mass within a two year period (*The Latin Mass*, Spring 1995, at 20). At a time of plummeting Mass attendance and vocations, the Church can scarcely afford to ignore such an opportunity for revival.

This is not to say that the traditional Latin Mass should be restored as the principal Roman rite. The traditional rite does not compete with the new liturgy; it complements it. Permitting the traditional Latin Mass maintains a sense of continuity between past and present. It preserves the form of religion for which countless ancestors in Britain, Ireland, and other countries suffered death and persecution. It contributes to a better understanding of how our liturgy has evolved. The liturgy that prevailed from the time of Pope Gregory the Great until the Second Vatican Council is part of our Catholic heritage and cannot simply be ignored or discarded.

The traditional Latin Mass is not only part of our religious heritage; it is part of our cultural heritage. Much of the world's greatest art and music would be rendered meaningless without the old Mass. In 1971, over fifty of the world's most distinguished scholars, writers, historians, and artists living in Britain, only a handful of whom were Catholic, petitioned the Holy See "to allow the traditional Mass to survive, even though this survival took place side by side with other liturgical forms." The Mass, they wrote, "belongs to universal culture as well as to churchmen and formal Christians." Signatories of the appeal, which was "entirely ecumenical and non-political," included writers Agatha Christie, Robert Graves, Graham Greene, violinist Yehudi Menuhin, and actor Ralph Richardson.

The traditional Latin Mass also complements the contemporary Roman rite by continuing to emphasize doctrinal aspects of the Mass that are downplayed in the current rite — aspects such as the sacrificial nature of the Mass, and the priest's role *in persona Christi*. The use of Latin in the liturgy helps keep alive the Church's official language and maintains the universal, transnational, and multicultural character of our faith.

I believe that Archbishop Dimino will approve an indult Mass on Fort Bragg if you so request. It is, of course, for you as Post Chaplain to determine the time, place, and frequency of such a Mass. The Priestly Fraternity of Saint Peter provides the celebration of the traditional Latin Mass at the invitation of the local bishop.

It would be my privilege to discuss this request with you in person if you so desire.

Timothy B. Sullivan

Major, U.S. Army

A few weeks passed during which I was immersed in the change of command and the initial blast of activity that came with assuming XO duties. Then, nearly simultaneously, both Msgr. Spiers and Tim received responses to their individual letters:

> July 26, 1996
>
> Dear Monsignor:
>
> I wish to acknowledge your letter of 10 July 1996 addressed to Archbishop Dimino.
>
> It is the policy of this Archdiocese to permit the celebration of the Mass in the form prescribed in *Ecclesia Dei* only after a request is made for such a celebration by the Catholic chaplain of the installation.
>
> I want to thank you for all your good work in service to our people in the military. May God bless your apostolic labors.
>
> Yours in Christ,
>
> Rev. Msgr. Aloysius. R. Callaghan, S.T.L., J.C.D.
> Moderator of the Curia/Chancellor

Telephone: (301) 853-0400
Fax: (301) 853-2246

Post Office Box 4469
Washington, D.C. 20017-0469

July 26, 1996

Reverend Monsignor Edward F. Spiers
St. Joseph of the Pines, Inc.
590 Central Drive
Southern Pines, NC 28387-2212

Dear Monsignor:

    I wish to acknowledge your letter of 10 July 1996 addressed to Archbishop Dimino.

    It is the policy of this Archdiocese to permit the celebration of the Mass in the form prescribed in "Ecclesia Dei" only after a request is made for such a celebration by the Catholic chaplain of the installation.

    I want to thank you for all your good work in service to our people in the military. May God bless your apostolic labors.

Yours in Christ,

Rev. Msgr. Aloysius R. Callaghan, J.C.D., S.T.L.
Moderator of the Curia/Chancellor

The letter Tim received was much more curt. He read it to me over the telephone in an agitated voice. I asked him to provide me with a copy of it, but he never did; quite frankly I think he was embarrassed by it—and this seemed to be the intent of the letters we received from these chaplains. Tim had often commented that the reason traditional Catholics were having so many difficulties with the clergy is that such Catholics were always making demands, they're rude and they don't know how to ask for something politely. After receipt of this letter, I would never hear him say that again.

In the next few weeks I spoke with every Catholic chaplain at Fort Bragg individually, either over the telephone or in person.

One, Fr. Jerome Haberek, who was the Chaplain for USASOC (my own higher headquarters), had only one word for me: "No!" One would think that a chaplain would have a bit more consideration for a fellow Catholic in his own unit, especially in the Special Operations Command. We had risky jobs, and at any moment we could have been called out into harm's way and killed. He would not even discuss it. All he would say was "No!"

Another priest took the time to engage in discussion. Fr. Patrick Healy seemed more amenable to the idea initially, then he bared his soul. At some point in the conversation on the phone, he snapped: "'The pope says...' 'The pope says...' I'm *tired* of hearing about what the pope says!"

Then, there was Fr. Czech. He took the time to meet with me; we had lunch together, and talked about the liturgical crisis in the Church. He knew exactly where I was coming from. He even agreed. "Oh, yes, the Latin Mass is more Catholic," he confided.

I couldn't believe my ears. "Why?"

"Because it's more universal. 'Catholic' means universal." He understood that the original intent of the Second Vatican Council was that the Mass was to remain, at least partly, in Latin, and that much of what had taken place with the new liturgy had not been intended by the Church.

Now this was a change. Perhaps I had found a priest who would be able to help. But Fr. Czech was only a Captain. The senior chaplain held the rank of Colonel, and in fact, all the other priests outranked Fr. Czech. But he was a priest, and he agreed to talk with Fr. Marceaux. He would meet with him as soon as possible; if it was not possible to start having a weekly Mass according to the 1962 Missal under the provisions of *Ecclesia Dei*, at least we could have a Latin Mass in the new rite. Under the provisions of Canon Law any priest could do that.

We parted and we didn't speak again for about a week. When I next saw him, he was quite disturbed. Fr. Marceaux had warned him in the sternest terms possible that he was to have nothing to do with me. Suddenly concerned about his military career, which he would soon lose anyway, he offered no resistance to this abuse of authority.

Then there was Fr. Eric Albertson. Fr. Albertson was also only a captain, but he was known for his orthodoxy. He had taken the unusual step of going through the Army's difficult Ranger School as well as Airborne School. A priest going through Ranger School would clearly come to a much better understanding of the life of an infantry soldier. One had to admire him for enduring the two-month trial.

Yet I could never really speak with him. Every time I tried, the strangest situation would emerge. Every time! I tried to discuss some possibilities for improving the liturgy for soldiers assigned to Fort Bragg; I tried to talk about Church documents, or the erroneous implementation of Vatican II. I really tried to discuss serious issues with him. I was concerned about the massive apostasy, the feminization of the liturgy, the bizarre ideologies of the left that had crept into our worship, and the terrible loss of souls. I, the soldier, wanted to talk about these things. And Fr. Albertson, the priest, would invariably want to shift the conversation to Ranger Tabs, Airborne wings, ribbons, and other trinkets that go on the military uniform. He wanted to talk about the most trivial of things that military people would want to talk about during their first few years in the Army. I was concerned about the salvation of souls, and he seemed more interested in military toys. Eventually I came to understand that, for whatever reason — whether ambition, a desire to protect his career, a lack of interest, or a warning he had received from his superior — he would make no waves. It was clear that he would never help us.

I went to Mass once when Fr. Albertson was scheduled at his chapel, which was the 82nd Airborne Division Chapel. Yes, he was known for his "orthodoxy," and it was clear that he upheld Catholic doctrine in his sermon, but sure enough — he could not avoid the feminization of the liturgy. Undoubtedly, he was unable to do anything about it. There were female altar servers, the music was vapid, and the unpleasant modernist ambiance left a tremendous spiritual void I had noticed whenever I attended the new Mass. Nothing here lifted the heart to God. When Mass ended, a gentleman about my age sitting in front of me shook his head and said, "Mass has become so politically correct." His irate wife snapped, "Honey, can you please keep your opinion to yourself!"

To my infinite regret, I was too stunned to tap him on the shoulder and initiate a conversation about finding a better way.

At this point certain things had become very clear to me. It was clear that the current state of the liturgy did not reflect the desires of the "Council Fathers" of Vatican II. It was also clear that each one of the Chaplains at Fort Bragg either:

1. Knew that there was a problem, but didn't have the courage or ability to do anything about it;

2. Was not intelligent enough to recognize that there was a problem, or

3. Was part of the problem, desiring to drive a wedge between the U.S. Catholics and the Vatican.

It was also clear that the archbishop of the Military Archdiocese would do nothing to help us.

Here the problem began to take on a more serious aspect, because military personnel, of all people, need to be exposed to good Catholic doctrine. The U.S. Armed Forces, capable of inflicting incalculable damage on the rest of the world, should be led by officers and senior enlisted men well versed in the Just War criteria and have a genuine, deep, *Catholic* respect for human life. The more disturbing thought that I began to consider was the impact of military Catholics at Fort Bragg, the heart and soul of the military fighting force, being turned against the Catholic Church. It seemed clear to me that the eventual outcome of letting this situation continue would be extremely harmful.

I continued to hold meetings with others who shared a desire for the Latin Mass; there were a couple of young Field Artillery officers from New York; both were in the same battalion. One was a lieutenant; the other was a captain who commanded a company. There were a couple of JAG officers by now, several other officers of various branches, two or three other families, and a few enlisted personnel. We circulated a petition and all of us signed it.

I presented the petition to Fr. Marceaux, in hopes that he would somehow be moved by the fact that the request was coming from several officers and senior enlisted men. He was quite frank in his response, now. He would *never* allow a Latin Mass of any type on this installation if he was the senior chaplain. He didn't care if there were 500 or 1,000 signatures. Since the bishop of Raleigh had not allowed it, he, Fr. Marceaux could not allow it. Not under any circumstances. Ever.

"Why?"

"Because, since it's not allowed in the Diocese of Raleigh, all these people from around the area who don't agree with the bishop of Raleigh would be coming onto the military installation to go to Mass. People who have nothing to do with the Army would be here for Mass."

"So what? It's an open installation. And it would be good to have a lot of people attending Mass."

"They would be running away from the churches of the Diocese of Raleigh. They need to stay in them and accept Vatican II." This wasn't the time to argue about Vatican II. I happened to know that our Maronite priest was looking for an opportunity to expand and to establish a second Maronite Catholic parish in the area somewhere. No small number of Catholics from Fort Bragg were already members of Archangel Michael, so perhaps he could have a Maronite liturgy on the installation. After all, the Holy Father had recently released a document entitled *Orientale Lumen*, encouraging us all to come to an appreciation of the Eastern rites.

"Would it be possible for our Maronite priest to come here, maybe once a month, and we could have Maronite Divine Liturgy on the installation?"

"No, that would not be appropriate."

"Why not?"

"It would divide the Church."

"Eastern rites are *part of* the Church!"

"No, I'm not going to allow the *Marionites* to use our chapel."

Marionites? Had he mispronounced it by mistake or did he not even know who they were? Then I remembered that I also didn't know anything about the Eastern Catholic churches a couple of years ago. It seemed acceptable for a lay person to be ignorant, but a priest? We stared at one another for a moment. Then, suddenly, he was no longer in command of the conversation; he looked somehow vulnerable.... now there seemed to be an opening for an actual conversation. He became inquisitive:

"Do they say their prayers in Latin, too?"

"No, but some of the prayers are in Syriac. It's very close to Aramaic, an ancient language Christ spoke in his daily life." He seemed interested in what I was saying for once, so I continued. "The Maronites are one of the many Eastern rites of the Catholic Church, so they have their own liturgy. They also have their own liturgical calendar, with different feast days — sometimes including Saints most westerners haven't heard of."

He was still listening! I was on a roll. "They have their own code of canon law, liturgical calendar, and their own hierarchy of bishops. For example, the Maronites here in Fayetteville aren't part of the Diocese of Raleigh. They come under the Eparchy of Saint Maron, and their bishop is in New York. An eparchy is the same thing as a diocese. Their Patriarch is in Lebanon. They have a very beautiful liturgy; and there are no girl altar boys. It's all quite traditional, but it's of a different tradition than what we're used to in the west. The forty days of Lent begin with Ash Monday instead of Ash Wednesday, for example..."

As I continued on, he became uneasy, then anger started to build again. This was obviously something new for him. This was something he wasn't in control of, and he didn't like it. There was nothing he could do about it. The Sonnier family remained safely out of his clutches, and there wasn't a thing he could do about it. He couldn't complain to the bishop of Raleigh, because the Maronites weren't under that jurisdiction. He couldn't stop us from going to Archangel Michael. He couldn't force us to accept girl altar boys. The faithful were avoiding the Novus Ordo by attending Mass with the Eastern Catholics, and he couldn't stop them!

Unfortunately, I was slow to understand that what I was saying angered him, and I was still rambling on with all the wonderful things about the Maronites and the Eastern Catholic Churches. I thought it might appeal to his sense of "diversity." "There are quite a few people from Fort Bragg at Archangel Michael..."

"How many?"

"Oh, twenty, thirty. Maybe more. It depends on whether you want to count contractors who work here, for example."

In my mistaken belief that he was impressed by this number, I took the opportunity to ask: "Don't you think it would be appropriate to invite the priest to offer Mass here in one of your chapels, perhaps once a month or something?"

"No, it would be divisive. I think our meeting is over," he hissed, straining, but unable to conceal his anger any longer.

I reported the sad news back to the Maronite priest the following weekend. He was unconcerned, as he had already been looking into the possibility of opening a church in Raleigh instead. It looked much more promising there. If Fort Bragg didn't want him, he would start an apostolate in Raleigh.[3]

Hoping to convince the bishop of Raleigh of the need to yield a bit on this issue, due to the difficulties he was causing for us, I sent him another letter in October explaining that the path he had advised me to take had failed due to his own policy. In my initial contact with him he had referred me to the Military Archdiocese; now the Military Archdiocese was pointing to his own policy as a reason for refusing us a Latin Mass. He did not respond.

Through the Latin Liturgy Association newsletter, I discovered the presence of another priest in the region who was favorable to the idea. Fr. John J. Nicola was a refugee from the Archdiocese of Chicago who lived in Southern Pines with his mother. For as long as the Archdiocese of Chicago was run by Cardinal Bernadin, it had been clear that loyal, orthodox priests were not welcome there. Fr. Nicola had moved to North Carolina to stay with his elderly mother. Meanwhile, he had some limited faculties from the Diocese of Raleigh, and he taught at a private school in the area. Fr. Nicola offered to help by offering Mass in the new rite, in Latin, on Sunday afternoons at a certain hour. We were quite grateful to Fr. Nicola. As it turned out, he was already offering Mass on Sundays in a private chapel in his house anyway, so it would be no difficulty for him.

We began making plans, but our group had grown larger, and somehow the word got back to Fr. Marceaux that we were going to have a Latin Novus Ordo Mass off base. He arranged a meeting with Fr. Nicola and told him to have nothing to do with us. Embedded

---

[3] At that time he began a "St. Sharbel Mission" in Raleigh, which is now known as St. Sharbel Church.

in the conversation, somehow, was a veiled threat of "excommunication," which Fr. Nicola reported back to me. He urged me to quit circulating the petition, and accept the current situation.

"Can't we come to Mass at your private chapel?" I asked.

No, it couldn't be permitted. He was a guest in the Diocese of Raleigh and was only allowed to offer private Mass for the sick. Bishop Gossman wanted every Catholic registered in a Novus Ordo parish, and he was afraid that Fr. Nicola's Mass would take people away from his agenda. It probably would have.

"Why should I stop circulating the petition and submitting requests?" I asked Fr. Nicola.

"Because Fr. Marceaux may have to take some disciplinary action, like suspend or excommunicate you, and then the Church would be weaker."

Now this was clearly abuse of authority. Any priest can offer the New Mass in Latin, and any layman may be present. Given the faulty translations of the new Missal that were in force at the time — a problem that would not be amended for many years — and given the plain language of *Sacrosanctum Concilium*, praying the New Mass in Latin would have been something any priest could do without permission. Furthermore, there was no basis for any kind of threat of excommunication. However, Fr. Marceaux clearly found the existence of a small but loyal group of Catholics so frightening that he would go way beyond his authority as a priest to make these intimidating threats.

In a few days I discovered it was a bluff. By now I could count on a larger network of Catholics for advice. In St. Louis, a gentleman by the name of Fred Haenel was organizing "*Una Voce America*" as a U.S. extension of an international organization that had been around since before the end of the Second Vatican Council (it was founded in 1964, to be precise). *Una Voce* had worked to ensure that traditional Catholic practice was not destroyed by the bizarre winds that belched out across the world during the 1960s. They had been extremely effective in the preservation of the old rite, and Archbishop Bugnini had complained about them bitterly whenever he had the chance. Over the previous months we had registered a chapter as "*Una Voce* Fort Bragg."

Through my contacts with this organization, I found the "Saint Joseph Foundation," a canon law society in San Antonio, Texas.

I sent them copies of the documents exchanged, and then one afternoon spent nearly two hours on the telephone with Chuck Wilson, a convert and a Catholic loyalist. He understood the situation perfectly. In fact, he had heard countless stories like ours. Priests and bishops caught up in ideological agendas hated Catholics who preferred the old Mass. They would abuse their authority, mislead and deceive, or drive out the faithful to have their way. In Hawaii a bishop had gone so far as to "excommunicate" people in a situation similar to ours. The "excommunication" was overturned by the Vatican, and that bishop soon retired. The bad news was that this kind of situation was going on everywhere; loyal Catholics like us were suffering the same abuse from coast to coast. The good news was that we could not be excommunicated.

Thank God for the St. Joseph Foundation. I made it a point to include a donation to them in our family budget. Unfortunately, Canon Law is easy to abuse by canon lawyers with ideological agendas. Chuck's office already knew something of Fr. Marceaux but they wouldn't say much about him. Anyway, there was no canonical provision that gave us an absolute right to our old liturgy. All Chuck could do was to allay my concern that I could somehow be subject to canonical penalty; I had every right to petition as I was doing. He gave me some suggestions for exploring other avenues with the Commanding General of Fort Bragg, or the Department of the Army Chief of Chaplains.

Why not?

My wife and I co-signed a letter to the 18th Airborne Corps commander, Lieutenant General Jack Keane. Being a Catholic and a graduate of the Jesuit Fordham University as well as a "three-star" general, surely, he might be moved to intervene. He was only a lay Catholic like us, but certainly he couldn't have reached the rank of general without maneuvering through some complex situations.

What I had yet to discover was that doing the right thing is not only a matter of intellect. A Catholic can understand the present crisis perfectly well yet lack the moral courage to take some small but significant step that will place them on the side of the good (in this case the Church) and against the prevailing foul winds. Lieutenant General Keane's response to our letter indicated that he had not even taken a few seconds to try to understand what we were asking of him:

DEPARTMENT OF THE ARMY
HEADQUARTERS, XVIII AIRBORNE CORPS AND FORT BRAGG
FORT BRAGG, NORTH CAROLINA 28307-5000

REPLY TO
ATTENTION OF: February 4, 1997

Office of the Corps Chaplain

Major and Mrs. David Sonnier
105 Frazier Court
Fort Bragg, North Carolina 28307

Dear Major and Mrs. Sonnier:

Thank you for your letter concerning your disappointment over the decision not to recommend your petition for the celebration of the traditional Latin Mass at Fort Bragg. I recognize that it was not the decision you sought, but, I feel Monsignor Marceaux was acting in the best interest of the Fort Bragg Catholic Community when he rendered to the Military Archbishop his judgment to deny.

Monsignor Marceaux used the teachings of the 1970 Roman Missal, which suppressed the traditional Latin Mass, as the basis for his judgment. It states the Pope's intention that Roman Catholics receive the new missal as a help toward strengthening their unity in a diversity of vernacular languages as they celebrate mass. It allowed exceptions for the elderly and for ill priests, with the understanding that they increase their knowledge of the new liturgy and accept the teachings of the Second Vatican Council. After meeting with you on several occasions, Monsignor Marceaux did not feel your request fell within the parameters of the Pope's guidelines.

Catholic priests serving on military installations receive their guidance from the Military Archbishop. He insures that when requests are evaluated, they are not in opposition to the policies of the local diocese. The Bishop of Diocese of Raleigh, which encompasses Fort Bragg, does not allow for the celebration of the traditional Latin Mass in any of the parishes, to include Fayetteville.

I understand you have a schedule of traditional Latin Mass celebrations in the United States, and I encourage you to pursue these as a means of meeting your personal religious needs.

Sincerely,

John M. Keane
Lieutenant General, U.S. Army
Commanding Officer

February 4, 1997

Dear Major and Mrs. Sonnier:

Thank you for your letter concerning your disappointment over the decision not to recommend your petition for the celebration of the traditional Latin Mass at Fort Bragg. I recognize that it was not the decision you sought, but I feel Monsignor Marceaux was acting in the best interest of the Fort Bragg Catholic Community when he rendered to the Military Archbishop his judgment to deny.

Monsignor Marceaux used the teachings of the 1970 Roman Missal, which suppressed the traditional Latin Mass, as the basis for his judgment. It states the Pope's intention that Roman Catholics receive the new missal as a help toward strengthening their unity in a diversity of vernacular languages as they celebrate mass. It allowed exceptions for the elderly and for ill priests, with the understanding that they increase their knowledge of the new liturgy and accept the teachings of the Second Vatican Council. After meeting with you on several occasions, Monsignor Marceaux did not feel your request fell within the parameters of the Pope's guidelines.

Catholic priests serving on military installations receive their guidance from the Military Archbishop. He ensures that when requests are evaluated, they are not in opposition to the policies of the local diocese. The bishop of the Diocese of Raleigh, which encompasses Fort Bragg, does not allow for the celebration of the traditional Latin Mass in any of the parishes, to include Fayetteville.

I understand that you have a schedule of traditional Latin Mass celebrations in the United States, and I encourage you to pursue these as a means of meeting your personal religious needs.

Sincerely,

John M. Keane
Lieutenant General, U.S. Army
Commanding General

Did he really expect that we could drive to Virginia or Atlanta to attend the nearest "authorized" Latin Mass? Our request would have been very easy to accommodate, and the number of people who had signed our petition, by now, was greater than the number attending the Episcopal service on Fort Bragg. I knew that because the Episcopal priest worked for me as battalion Chaplain.

Furthermore, Lt. Gen. Keane had mentioned nothing about recent developments, such as *Ecclesia Dei*, in his letter. It seemed best to give him the benefit of the doubt for the moment. I surmised that Fr. Marceaux had hoodwinked him with some typical contemporary talking points about Vatican II and then written the letter Lt. Gen. Keane had signed.

I decided to try ecclesiastical channels again. Having appealed to both the Diocese of Raleigh and the Military Archdiocese to no avail, we decided that it was time to write to someone in the Vatican. I was short on time, because of an upcoming deployment, so 1Lt. Jeff McGowan, a young Field Artillery officer from New York, agreed to draft it for me and I would sign my name to it. He obtained the address of His Eminence Cardinal Felici and wrote a very nice letter outlining the curious responses we had received from everyone in the hierarchy, both within the Military Archdiocese and the Diocese of Raleigh. Cardinal Felici was the prefect of the *Ecclesia Dei* Commission, the office in the Vatican responsible for issues involving the old rite. Would Cardinal Felici be willing to write to Lt. Gen. Keane and explain the situation to him? Perhaps hearing from someone in the Vatican would make a difference.

We mailed the letter on February 27, and the response, dated April 23, arrived several weeks after that date. A few other things happened in the meantime.

First, the Catholic women at Fort Bragg elected my wife to be the president of the MCCW, which meant that Fr. Marceaux had to be much more careful about what he said about us in public. In fact, we heard rumors that he would be leaving soon, and we began to look forward to meeting his replacement. Hopefully it would be someone more respectful toward our aspirations and the guidance from the pope.

Then, in late March of 1997, I received a very encouraging letter from none other than Michael Davies, the President of *Una Voce* International. I contacted him by telephone to thank him for the letter and asked if I could make copies of it and distribute them. Of course, he didn't mind if I made copies of his letter, and I knew who to send them to. The letter is reproduced here:

Foederatio Internationalis Una Voce

14 March, 1997

Dear Major Sonnier,

I am very sorry to learn that members of the Fort Bragg Chapter of *Una Voce* are having difficulty in achieving their rightful aspirations for regular access to the celebrations of Mass according to the 1962 Roman Missal. As a former regular soldier, I am very surprised that this should be the case in a military establishment where, I would have imagined, loyalty and obedience to the wishes of a lawful superior could be taken for granted. I can only imagine that this is because the Military Chaplaincy is not clear about the position of the Holy Father in this matter.

On 2nd July 1988 His Holiness Pope John Paul II promulgated his *Motu Proprio "Ecclesia Dei"* in which he expressed his will to guarantee respect for the rightful aspirations of those attached to the Latin liturgical tradition, and in order to achieve this aim he established the Pontifical Commission *Ecclesia Dei*. On 18th October 1988, Pope John Paul II granted to Cardinal Mayer special faculties to facilitate the working of the Commission, the first of which reads:

"The faculty of granting to all who seek it (*omnibus id petentibus*) the use of the Roman Missal according to the 1962 edition, and according to the norms proposed in December 1986, by the Commission of Cardinals constituted for this very purpose, the diocesan bishop having been informed."

It is important to note that this faculty refers to *all* who seek the 1962 Missal. As President of the *Ecclesia Dei* Commission, Cardinal Mayer provided an authoritative interpretation of the *Motu proprio* in a letter to the Bishops of the U.S. dated 20 March 1991. In this letter he explained that the Holy Father:

"...addressing himself 'to all those Catholic faithful who feel attached to some previous liturgical and disciplinary forms of the Latin tradition,' *and not just to former adherents of Archbishop Lefebvre*, he expressed his will to 'guarantee respect for their *rightful* aspirations' (no. 5.c). In order to provide for these *legitimate* desires of the faithful he established this Pontifical Commission and indicated his mind with regard to its primary task by stating:

"'Respect must *everywhere* be shown for the feelings of all those who are attached to the Latin liturgical tradition, by a *wide and generous* application of the directives already issued some time ago by the Apostolic See for the use of the Roman Missal according to the typical edition (no. 6,c).'

"Consequently, Your Excellency, we wish to encourage you to facilitate the proper and reverent celebration of the liturgical rites according to the Roman Missal of 1962 *wherever* there is a genuine desire for this on the part of the faithful." (Our emphasis)

You will note that Cardinal Mayer, quoting directly from *Ecclesia Dei*, refers to "*rightful* aspirations" and "*legitimate* desires," and adds that "it would seem unnecessary, even unduly painful, to impose further restrictions upon those who wish to attend such celebrations." Many bishops have responded to this admonition in a very positive manner, and in most dioceses where this has been the case there are no longer any restrictions attached to the celebration of the Tridentine Mass. Hundreds of such Masses are now celebrated in parish churches, in Europe, Australia, New Zealand, Canada, and throughout the U.S. They are scheduled public Masses fulfilling the Sunday obligation which any member of the faithful is welcome to attend.

I understand that in the U.S. some members of the clergy at all levels have the mistaken impression that the use of the 1962 Missal is restricted to older people, those who had become used to the 1962 Missal before the publication of the new one in 1970. When I last met Cardinal Mayer in November 1996, he assured me that this was most certainly not the case, and he was kind enough to confirm this in writing for the members of our Federation on 31 January 1997, stating:

"In recent times it has been affirmed that the allowances given for the celebration of the 'Tridentine Mass' have been granted with the provision that only those who were familiar with the preceding forms of the liturgy would be allowed to benefit from those concessions.

"During my time as President of the Pontifical Commission *Ecclesia Dei* such limitation was never mentioned by the authorities involved. In this regard it should be mentioned that the Commission has used the faculty of erecting religious institutes which would benefit from the use of the Roman Missal of 1962 and other liturgical books in force at that time. Evidently it was understood that young recruits would be admitted to such communities and would benefit from all the concessions made to them. Hence one cannot speak of an age limit."

Priests from two of the religious institutes mentioned by His Eminence are now pursuing their apostolate in about sixteen American dioceses with the approval and encouragement of the diocesan bishops — the Priestly Fraternity of St Peter and the Institute of Christ the King. The first now has an American seminary which is swamped with vocations from young men many of whom were not born when the 1970 Missal was published. Perhaps a visit to Fort Bragg of a representative from one of these Institutes might help to clarify the situation for your chaplain of whose good will I am confident you can be certain.

I am sure that the following message dated 25 July 1996 from Cardinal Ratzinger to our Federation will make the position of the Holy See very clear:

"The International Una Voce Federation has played an important role in supporting the use of the 1962 edition of the Roman Missal in obedience to the directives of the Holy See. For this valuable service I express my gratitude to the members of the Federation and extend my blessing."

I enclose two reports from the official Catholic press in England describing the launching in England of a movement to promote the 1962 Missal initiated entirely by young people.

If I can be of any further help to you, please do not hesitate to contact me.

I wish you and the members of your Chapter every grace and blessing for the coming holy season of Easter.

Yours sincerely in Domino,
Michael Davies,
President.

Foederatio Internationalis Una Voce

24 Cromwell Avenue, BROMLEY, Kent, BR2 9AQ, England

'phone 44 181 402 2248   Fax 44 181 289 6377

14 March 1997

Major David Sonnier
105 Frazier Court
Fort Bragg
North Carolina 28307
U.S.A.

Dear Major Sonnier,

I am very sorry to learn that members of the Fort Bragg Chapter of *Una Voce* are having difficulty in achieving their rightful aspirations for regular access to the celebrations of Mass according to the 1962 Roman Missal. As a former regular solider, I am very surprised that this should be the case in a military establishment where, I would have imagined, loyalty and obedience to the wishes of a lawful superior could be taken for granted. I can only imagine that this is because the Military Chaplaincy is not clear about the position of the Holy Father in this matter.

On 2nd July 1988 His Holiness Pope John Paul II promulgated his *Motu Proprio "Ecclesia Dei"* in which he expressed his will to guarantee respect for the rightful aspirations of those attached to the Latin liturgical tradition, and in order to achieve this aim he established the Pontifical Commission *Ecclesia Dei*. On 18th October 1988, Pope John Paul II granted to Cardinal Mayer special faculties to facilitate the working of the Commission, the first of which reads:

> The faculty of granting to all who seek it (*omnibus id petentibus*) the use of the Roman Missal according to the 1962 edition, and according to the norms proposed in December 1986, by the Commission of Cardinals constituted for this very purpose, the diocesan bishop having been informed.

It is important to note that this faculty refers to *all* who seek the 1962 Missal. As President of the *Ecclesia Dei* Commission, Cardinal Mayer provided an authoritative interpretation of the *Motu proprio*. In a letter to the Bishops of the U.S.A. dated 20 March 1991. In this letter he explained that the Holy Father:

> ...addressing himself "to all those Catholic faithful who feel attached to some previous liturgical and disciplinary forms of the Latin tradition", *and not just to former adherents of Archbishop Lefebvre*, he expressed his will to "guarantee respect for their *rightful aspirations*" (no. 5. c). In order to provide for these *legitimate desires* of the faithful he established this Pontifical Commission and indicated his mind with regard to its primary task by stating:

> respect must everywhere be shown for the feelings *of all those* who
> are attached to the Latin liturgical tradition, by a wide and
> generous application of the directives already issued some time ago
> by the Apostolic See for the use of the Roman Missal according to
> the typical edition (no. 6, c).

> Consequently, Your Excellency, *we wish to encourage you to facilitate the
> proper and reverent celebration of the liturgical rites according to the
> Roman Missal of 1962 wherever there is a genuine desire for this on the
> part of the faithful.* (Our emphasis)

You will note that Cardinal Mayer, quoting directly from *Ecclesia
Dei*, refers to "*rightful* aspirations" and "*legitimate* desires", and
adds that "it would seem unnecessary, even unduly painful, to
impose further restrictions upon those who wish to attend such
celebrations". Many bishops have responded to this admonition in a
very positive manner, and in most dioceses where this has been the
case there are no longer any restrictions attached to the
celebration of the Tridentine Mass.   Hundreds of such Masses are
now celebrated in parish churches, in Europe, Australia, New
Zealand, Canada, and throughout the U.S.A.   They are scheduled
public Masses fulfilling the Sunday obligation which any member of
the faithful is welcome to attend.

I understand that in the U.S.A. some members of the clergy at all
levels have the mistaken impression that the use of the 1962 Missal
is restricted to older people, those who had become used to the
1962 Missal before the publication of the new one in 1970.   When I
last met Cardinal Mayer in November 1996 he assured me that this
was most certainly not the case, and he was kind enough to confirm
this in writing for the members of our Federation on 31 January
1997, stating:

> In recent times it has been affirmed that the allowances given for the
> celebration of the "Tridentine Mass" have been granted with the provision
> that only those who were familiar with the preceding forms of the Roman
> liturgy would be allowed to benefit from those concessions.

> During my time as President of the Pontifical Commission "Ecclesia Dei"
> such limitation was never mentioned by the authorities involved. In this
> regard it should be mentioned that the Commission has used the faculty of
> erecting religious institutes which would benefit from the use of the
> Roman Missal of 1962 and the other liturgical books in force at that time.
> Evidently it was understood that young recruits would be admitted to such
> communities and would benefit from all the concessions made to them.
> Hence one cannot speak of an age limit.

Priests from two of the religious institutes mentioned by His
Eminence are now pursuing their apostolate in about sixteen
American dioceses with the approval and encouragement of the
diocesan bishops—the Priestly Fraternity of St Peter and the
Institute of Christ the King.   The first now has an American
seminary which is swamped with vocations from young men many of
whom were not born when the 1970 Missal was published.   Perhaps a
visit to Fort Bragg of a representative from one of these

2

Institutes might help to clarify the situation for your chaplain of whose good will I am confident you can be certain.

I am sure that the following message dated 25 July 1996 from Cardinal Ratzinger to our Federation will make the position of the Holy See very clear.

> The *International Una Voce Federation* has played an important role in supporting the use of the 1962 edition of the Roman Missal in obedience to the directives of the Holy See. For this valuable service I express my gratitude to the members of the Federation and extend my blessing.

I enclose two reports from the official Catholic press in England describing the launching in England of a movement to promote the 1962 Missal initiated entirely by young people.

If I can be of any further help to you please do not hesitate to contact me.

I wish you and the members of your Chapter every grace and blessing for the coming holy season of Easter.

Yours sincerely in Domino,

Michael Davies,
President.

The letter was accompanied by a photocopy of an article in *The Catholic Times*, dated March 9, 1997, entitled "A Thousand-plus Flock in for Tridentine Mass." The article describes the launching of *Centre International d'Etudes Liturgiques* (CIEL) in the UK, and a Votive High Mass of Our Lady at St. James Church, Spanish Place, London. Each of the chaplains at Fort Bragg, and the office of the Chief of Chaplains in the Pentagon, soon received a copy of this letter.

We also contacted Cardinal Alfons Maria Stickler about this time. One of our acquaintances knew Msgr. Ignacio Barreiro, who was in Rome finishing a doctorate in Thomistic theology. We contacted Msgr. Barreiro and asked him to inform Cardinal Stickler of our plight and see if he had any suggestions. Cardinal Stickler had been the Vatican Librarian and Archivist. He was retired by now, but in 1996 he had celebrated a traditional Latin Mass in St. Patrick's Cathedral in New York, creating a considerable stir in the media and sending shock waves through the ranks of the modernists. Msgr. Barreiro wrote back to us:

> Last Wednesday, April 9th, I received your letter with all the enclosures for Cardinal Stickler, which you had mailed on March 7th. I had been expecting for some time the arrival of your letter because Roger McCaffrey explained the situation to me last February. On Saturday April 12th, I had a meeting with Cardinal Stickler, and I delivered to him all the materials.
>
> The Cardinal...told me that he will study the papers and he will answer me. I hope to see him again by the end of the month....
>
> May the Lord grant you all His blessings in this Eastertide and may He assist you in your efforts to preserve the traditional Liturgy of the Church....

This flurry of letters also included one to the Department of the Army Chief of Chaplains, a two-star general. Each of the branches of the service has a similar position. Within the Army the office rotates on a two-year basis; a Catholic is assigned for two years, then a Protestant, then a Catholic again. Now the office was held by Major General Donald W. Shea, a Catholic. I wrote to him on April 7, 1997, and I explained the love we had acquired for the old Mass as well as the fact that it was not available at Fort Bragg or the surrounding area. Specifically, I asked him the following questions:

Two distinct positions, with respect to the Tridentine Mass, are articulated. Which is correct — that articulated by Mr. Davies or that articulated by Lt. Gen. Keane?

I understand that the Chaplains Corps is having extreme difficulties in obtaining priests for the chaplaincy. Given that the traditional orders in good standing with the Vatican (FSSP, and the Institute of Christ the King) are flourishing, at some point will you begin taking in their members as military chaplains?

Currently there is no Tridentine Mass anywhere in the Carolinas, yet continuation of my Special Forces career requires that I spend considerable time here at Fort Bragg. Is it the position of the Chaplains Corps that I should sacrifice my career (leave the service or transfer out of Fort Bragg) if I would like to have the Tridentine Mass available?

Currently the Fort Bragg Chapter of Una Voce is not allowed to meet in Fort Bragg's chapels, nor can we advertise in the bulletins. According to the appropriate Army Regulations, may a chaplain of a particular denomination deny an organization (in good standing with that same denomination) the right to announce and hold meetings?

While awaiting the answer to this letter, I had some time to think. Why was this so important to me? And why do men risk or give up their lives for their country in combat? Was there a relationship between these two questions? I had been pondering this thought for a while, and finally I was able to reach an answer.

Yes, I now concluded, there was. The reason I had been successful as a young military officer, despite my apathy about administration, was that I was unafraid. I had the same fear that others experienced when confronted with a dangerous situation, but the fear of combat or death during an operation could be easily overcome by making a good confession, knowing that there would be eternal reward for doing my best. "Greater love has no one than this, that he lay down his life for another" (John 15:18). But I had to be convinced that what I was laying down my life for was worth defending! Should one pay the ultimate price in the service of an Army whose Catholic priests show complete disregard for the highest authority of the Church? Could I defend a country whose Catholics appeared to be acting no differently? I spent some time discussing this issue with my boss, who, for a non-Catholic, seemed to have an extraordinary level of sympathy. By now I had come to admire and respect my battalion commander, Lt. Col. David Fridovich, better known as "Frido."

"Why did you come into the Special Forces, sir?" I asked him.

"Free coffee."

"Huh?"

"Free coffee. I always liked coffee. Well, *Liber* is free, and *Oppresso* is coffee, so *De Oppresso Liber*[4] means free coffee. Or so I thought. It wasn't until I had already joined Special Forces that I realized it was *De Oppresso Liber*, not *de espresso liber*, which did in fact change everything."

Passing the guidon to LTC David Fridovich during a change of command ceremony, August 1996 at Fort Bragg, NC, relinquishing command of B Company, 2nd Battalion, 3rd SF Group.

Frido seemed to understand the predicament we were in. He invited my family to his family's Passover Seder, and we learned about Orthodox Judaism. He read the scriptures and prayers in Hebrew, interjecting an explanation along the way as necessary. The Orthodox Jews had some issues to wrestle with that were like ours. I was curious about efforts to feminize Judaism. "Sir, what do you think of female rabbis?" I asked.

---

4 *De Oppresso Liber*, the Special Forces motto, literally means "from [being] an oppressed man [to being] a free one," but it is popularly taken to mean "to free from oppression" or "to free the oppressed."

"I don't know what you're talking about."

"You know…women who become…."

"I don't know what you're talking about!!"

He was sympathetic to our plight. His prayers in Hebrew, recited with his family, were as lonely as our family prayers in Latin. From that time on, I've often noticed that Catholics who lose their minds over seeing us praying in our prayer language, Latin, become speechless when you point out that Jewish people pray in Hebrew and many of the Orthodox pray in Church Slavonic. A prayer language is not at all unusual, but for some reason our civilization had become hostile to the use of the traditional prayer language of Western Christianity.

Eventually the response came back from the office of the Army Chief of Chaplains:

> 23 April 1997
>
> Dear Major Sonnier:
>
> This is in response to your letter of April 17, 1997 to Chaplain Shea concerning the lack of a Tridentine Mass being celebrated at Fort Bragg, North Carolina. Chaplain Shea has asked me to respond to you.
>
> The religious program at Ft Bragg is the commander's program. The letter you received from LTG Keene [*sic*], Commander of the XVIII Airborne Corps explains the situation very well. He must rely on the advice of his chaplains. The chaplains follow the guidelines issued by the Archdiocese for the Military.
>
> We in the Army Chaplaincy have no knowledge of the "traditional orders" you mentioned in your letter. The Archdiocese for the Military is the official endorsing agent for the Roman Catholic Church. We would have to receive an endorsement from them before we could even consider an application for the Military Chaplaincy.
>
> The Chaplain Corps is not taking any position on your career. You are the best manager of that part of your life. You must make your own career decisions.
>
> The use of the Chapel facilities at Fort Bragg are under the control of the installation commander. He, again, must rely on the judgement of his chaplains about the use of the chapels.

DEPARTMENT OF THE ARMY
OFFICE OF THE CHIEF OF CHAPLAINS
2700 ARMY PENTAGON
WASHINGTON DC 20310-2700

REPLY TO
ATTENTION OF

April 23, 1997

Major David L. Sonnier
105 Frazier Ct
Fort Bragg, NC 28307

Dear Major Sonnier:

This is in response to your letter of April 17, 1997 to Chaplain Shea concerning the lack of a Tridentine Mass being celebrated at Fort Bragg, North Carolina. Chaplain Shea has asked me to respond to you.

The religious program at Ft Bragg is the commander's program. The letter you received from LTG Keene, Commander of the XVIII Airborne Corps explains the situation very well. He must rely on the advise of his chaplains. The chaplains follow the guidelines issued by the Archdiocese for the Military.

We in the Army Chaplaincy have no knowledge of the "traditional orders" you mentioned in your letter. The Archdiocese for the Military is the official endorsing agent for the Roman Catholic Church. We would have to receive an endorsement from them before we could even consider an application for the Military Chaplaincy.

The Chaplain Corps is not taking any position on your career. You are the best manager of that part of your life. You must make your own career decisions.

The use of the Chapel facilities at Fort Bragg are under the control of the Installation commander. He, again, must rely on the judgement of his chaplains about the use of the chapels.

Please be assured that you and your family will be in my prayers. Please keep us in yours.

Sincerely,

John J. Kaising
Chaplain (Colonel) US Army]
Executive Officer

 Printed on   Recycled Paper

Please be assured that you and your family will be in my prayers. Please keep us in yours.

Sincerely,

John J. Kaising
Chaplain (Col.) U.S. Army
Executive Officer

Three weeks later, this chaplain was featured in the *Army Times*, boasting of his efforts on behalf of a soldier who desired his assistance in establishing a Wiccan "chapel service." It appears that it was of extreme importance to Fr. Kaising that he intervene with her commander to ensure that she could practice her "faith":

> As an example, Kaising explained how he once assisted a Wicca minister who needed to use knives in her worship service. "I had to go to her commander and explain the importance of the use of those knives and, ultimately, he accommodated her request."[5]

Fr. Kaising, a priest in the Roman Catholic Church, and the Executive Officer of the U.S. Army Chief of Chaplains, intervened on behalf of someone attempting to engage in witchcraft. Yet he would deny one of his own spiritual children the Mass that the pope himself had asked him to permit generously. The Mass that was then not to be found in the entire state of North Carolina except for in the chapels of the Society of Saint Pius X (SSPX). Why did he not offer to intervene in our situation as he had so generously done on behalf of a witch?

And I, a loyal Green Beret ready to risk my life for my country, would have remained willing to do so. But by now I was beginning to question whether I was serving in an Army that had the moral authority to make such a demand on my life. Was it prudent for me to be serving in such an army, as a father of a growing family? Where witches were welcome but the Latin Mass was not? What kind of Catholic could take such an army seriously?

Fr. (Col.) Kaising, who had intervened with a commander to ensure that some soldier would be provided with knives for a Wiccan service but wouldn't intervene on our behalf, would go on to become an auxiliary bishop in the Military Archdiocese.

I began looking into the history of the SSPX, which had been founded by Archbishop Lefebvre.

---

[5] *Army Times*, May 19, 1997, p. 14.

CHAPTER 9

THE SOCIETY OF SAINT PIUS X

I T WAS CLEAR THAT THE BISHOP OF THE DIOCESE of Raleigh and many other bishops throughout the world were actively suppressing the old Mass, contrary to the pope's clear plea for generosity. Several of the soldiers at Fort Bragg were committed to driving more than an hour away to attend Mass in a chapel with the Society of St. Pius X (SSPX). I always avoided doing so, thinking that it would be somehow best to remain in the "visible" structure of the Church and to work patiently with the "bishops and all those engaged in the pastoral ministry in the Church," as the pope had urged in *Ecclesia Dei.* But because there were SSPX chapels in Goldsboro and Raleigh, I at least considered the option.

I came late to the game. It had taken me an embarrassingly long time even to figure out that there was a problem, and I resolved to be careful in my search for a solution. For many years now, people had questioned the validity of the new Mass, based on the words of St. Pius V in the *De Defectibus* of the missal where he lays out the requirements for validity as well as defects that would result in an invalid sacrament:

> Defects on the part of the form may arise if anything is missing from the complete wording required for the act of consecrating. Now the words of the Consecration, which are the form of this Sacrament, are: *hoc est enim corpus meum,* and *hic est enim calix sanguinis mei, novi et aeterni testamenti: mysterium fidei: qui pro vobis et pro multis effundetur in remissionem peccatorum.* If the priest were to shorten or change the form of the consecration of the Body and the Blood, so that in the change of wording the words did not mean the same thing, he would not be achieving a valid Sacrament. If, on the other hand, he were to add or take away anything which did not change the meaning, the

> Sacrament would be valid, but he would be committing
> a grave sin.[1]

The words are: "This is my body"; "This is the chalice of my blood, of the new and eternal testament: the mystery of faith: which will be shed for you and for many unto the remission of sins." The substitution of the "for all" instead of "for many" that was commonly found in the Novus Ordo in 1997 prompted some to conclude that the formula of consecration had been improperly translated into English, rendering it invalid. They did not attempt to make this same case against the Novus Ordo Mass celebrated in Latin since the proper words of consecration are used in that case.[2]

It was not Catholics alone who protested against the abolition of the old rite. In 1971 more than fifty distinguished scholars, historians, writers, and artists living in Britain, most of them not Catholic, wrote to Pope Paul VI to appeal to him to protect the Latin liturgy from extinction. In response to this petition, Pope Paul VI ordered a document to be drawn up that made it possible for the old Mass to continue to be celebrated publicly in England (under rather restrictive conditions, it must be said). It is said he acted promptly upon seeing the name of one of his favorite writers among the list of names, namely, Agatha Christie's. This permission, granted in a letter to Cardinal Heenan, thus became known as the "Agatha Christie Indult." Tim had mentioned this historic petition in his excellent letter that we had circulated among the chaplains.[3]

In the postconciliar chaos and confusion a key player began to emerge. Marcel Lefebvre was born November 29, 1905, into a

---

[1] *De Defectibus*, no. 20.

[2] The Novus Ordo Missae formulas are somewhat different but convey the same essence: "Accípite et manducáte ex hoc omnes: hoc est enim Corpus meum, quod pro vobis tradétur" and "Accípite et bíbite ex eo omnes: hic est enim calix Sánguinis mei novi et ætérni testaménti, qui pro vobis et pro multis effundétur in remissiónem peccatórum. Hoc fácite in meam commemoratiónem." See Peter Kwasniewski, "Why the Omission of 'Mysterium Fidei' Does Not Invalidate the Consecration of the Wine," *New Liturgical Movement*, June 26, 2023, www.newliturgicalmovement.org/2023/06/why-omission-of-mysterium-fidei-does.html.

[3] See page 95 above. For a thorough account of this initial indult from Paul VI and the many other famous petitions that were published in defense of the Mass, see Joseph Shaw, ed., *The Latin Mass and the Intellectuals: Petitions to Save the Ancient Mass from 1966 to 2007* (Arouca Press, 2023).

devout Catholic family in French Flanders. He was one of eight children in his family, and five of the eight would eventually become priests or nuns. Lefebvre recognized his vocation at an early age, and was ordained September 21, 1929. He joined the Holy Ghost Fathers and spent some time as a missionary and a seminary professor. Eventually he was made a bishop, then later promoted to archbishop of Dakar (Senegal), and after demonstrating exceptional leadership qualities, he became archbishop delegate, the pope's representative, to all French Africa.

His accomplishments in Africa are nothing less than remarkable, and while they are a matter of historical record, the full impact of his work is seldom appreciated. Suffice it to say he laid the foundation for the conversion of a portion of the African continent.

By the time he returned to France, the year of the opening of Vatican II, he had spent nearly thirty years in Africa. He participated in all four sessions of the Council, serving as an important organizer among the conservative wing.[4] He was the archbishop of Tulle until 1968, when he resigned before the most radical changes were to go into effect. Shortly after his retirement he was persuaded by seminarians to come to their aid. Appalled by the lack of traditional seminary life and formation available to these men, whose only desire was to faithfully serve the Church, he decided to intervene despite his advanced age. He founded the Priestly Society of Saint Pius X, which was dedicated to preserving Catholic tradition and doctrine, and he acted as its first Superior General. This foundation came with the full approval of the Church. From that point on, he did all he could to be faithful to his episcopacy, traveling the world and encouraging Catholics to remain true to their faith and to uphold the traditions of the Church.

Tensions between the SSPX and the Vatican began to emerge from the time the exploding young order was founded. In 1974 two Apostolic Visitors were sent to inspect their seminary in Ecône, Switzerland. The archbishop felt the need to reply in a

---

4 Jerome Stridon, "The Coetus: Trad Godfathers at Vatican II," *OnePeterFive*, December 12, 2022, https://onepeterfive.com/coetus-trad-godfathers-vatican-ii/, and idem, "Trad Godfathers at Vatican II: Lefebvre on the Eve of the Council," *OnePeterFive*, January 24, 2025, https://onepeterfive.com/trad-godfathers-at-vatican-ii-lefebvre-on-the-eve-of-the-council/.

letter known as "The Declaration," since the two visitors caused considerable scandal: "These two Visitors from Rome considered it normal and inevitable that there should be married clergy; they did not believe there was an Immutable Truth and they also had doubts concerning the traditional concept of Our Lord's Resurrection...."[5]

The "Declaration" was nothing more than a firm statement of intent to adhere to the "age-old magisterium, in the conviction that we can thus do no greater service to the holy Catholic Church, to the Sovereign Pontiff, and to future generations."[6] He condemned "neo-Modernist and neo-Protestant tendencies, such as were clearly manifested during the Second Vatican Council, and after the Council in all the resulting reforms," which he credited as having contributed to the "demolition of the Church, to the ruin of the priesthood, to the destruction of the Holy Sacrifice and the Sacraments, to the disappearance of religious life, and to naturalistic and Teilhardian teaching in universities, seminaries, and catechetics, a teaching born of Liberalism and Protestantism many times condemned by the solemn magisterium of the Church."[7]

As the world sank into darkness and apostasy with widespread abandonment of vocations, plummeting Mass attendance, doctrinal confusion, and liturgical chaos, Archbishop Lefebvre quickly filled the void. He was surprised by his sudden popularity and the demands placed on his time, but he fulfilled his duties humbly, rejecting the notion that he was doing something special. "I must dispel a misunderstanding so as not to have to return to it," he said. "I am not the head of a movement, even less the head of a particular church. I am not, as they never stop writing, 'the leader of the traditionalists.'"[8]

Of course, the widespread admiration and respect for him was not shared by many in the Church hierarchy. This popular support amounted to a stunning rejection of the implementation of Vatican II, and insulted those working for radical change. By following Archbishop Lefebvre, these people, even if they didn't

---

[5] Fr. François Laisney, *Archbishop Lefebvre and the Vatican* (Angelus Press, 1998), 8.
[6] Laisney, 9.
[7] Laisney, 9.
[8] Archbishop Marcel Lefebvre, *Open Letter to Confused Catholics* (Angelus Press, 1986), 7.

say a word, made a clear statement that they had no interest in Vatican II reforms. They chose the preconciliar version of Catholicism, even if traditional Catholic life was more difficult and more demanding. His fellow bishops criticized him bitterly, and the French news agencies followed his every step.

As darkness settled over the Catholic Church, he saw the way out, yet he never credited himself as somehow having superior insight into the emerging crisis:

> On the 29th of August 1976, the whole of France was excited on hearing that I was going to say Mass at Lille. What was so extraordinary about a bishop celebrating the Holy Sacrifice? I had to preach before a panoply of microphones and each of my remarks was greeted as if it were a striking declaration. Yet what did I say beyond what any other bishop could have said? There lies the key to the enigma: the other bishops had been for a number of years no longer saying the same things....[9]

He did not criticize the other bishops for losing their faith, but simply credited the circumstances which the others lacked: "How have all these bishops been able to metamorphose themselves in this manner? I can see only one explanation: they were always in France, and they let themselves become gradually infected. In Africa I was protected."[10]

The archbishop attended the Second Vatican Council and participated in forming the document on the liturgy, even signing *Sacrosanctum Concilium*. But what he signed his name to has already been described. The use of Latin was to be preserved in the Latin rites. The treasury of sacred music was to be preserved and fostered with great care. The Church, recognizing Gregorian chant as being especially suited to the Roman liturgy, was to give it chief place in liturgical services. Obviously, none of these clauses was taken seriously.

Additionally, all lawfully acknowledged rites were to be treated as having equal authority and dignity, and were to be preserved in the future, and there were to be no innovations unless absolutely necessary for the good of the Church. Any new forms adopted would have to grow organically from existing forms. While many are fond of pointing out that "Archbishop Lefebvre

---

[9] Ibid.
[10] Lefebvre, 8.

signed the document on the liturgy," it is abundantly clear that what he signed was not what was implemented.[11]

Much more was written and said about Archbishop Lefebvre than he wrote or said about himself. He was "disobedient," he was "rebellious," and most grievously, he "rejected Vatican II." Yet, he was obedient to a fault, submissive to the pope as long as he echoed the timeless wisdom of the Catholic Church and rejected false or modern interpretations of the Catholic Faith:

> Obedience is a serious matter; to remain united to the Church's Magisterium and particularly to the Supreme Pontiff is one of the conditions of salvation. We are deeply aware of this, and nobody is more attached to the present reigning successor of Peter, or has been more attached to his predecessors, than we are. We are attached to the Pope for as long as he echoes the apostolic traditions and the teachings of all his predecessors...[12]

Pope John Paul II, in *Crossing the Threshold of Hope*, later admitted that there were false interpretations of Vatican II: "We feel the need to speak about the Council in order to interpret it correctly and defend it from tendentious interpretations. Such interpretations do in fact exist..."[13] Lefebvre was unique in that he rejected these biased interpretations, as all bishops should have done. He didn't only feel the need to speak, he actually spoke. He didn't merely wish that something could be done about it; he actually did something about it.

It has been argued that he was rebellious in not going along with the new Mass. Yet in retrospect, we can see that many bishops should have refused to go along with it, given its abrupt departure from the directives of the Second Vatican Council and its obvious rupture from bimillennial tradition. Furthermore, there was no absolute requirement that it be accepted by anyone. Archbishop Lefebvre studied the situation and arrived at the conclusion, later to be confirmed as true, that the Old Mass had never been abrogated.

---

[11] For a crystal-clear example of how the Council Fathers were deceived, see Peter Kwasniewski, "The Lie That Was Told to Over 2,000 Council Fathers at Vatican II," *New Liturgical Movement*, May 27, 2024, www.newliturgicalmovement.org/2024/05/the-lie-that-was-told-to-over-2000.html.

[12] Lefebvre, *Open Letter*, 129.

[13] John Paul II, *Crossing the Threshold of Hope*, 157.

> This is why we hold firmly to the Sacrifice of the Mass.
> And we are convinced that our Holy Father, the Pope,
> has not forbidden it and that no one can ever forbid the
> celebration of the Mass of All Time (*Messe de Toujours*).
> Moreover, Pope St. Pius V proclaimed in a solemn and
> definitive manner that, whatever might happen in the
> future, no one might ever prevent a priest from cele-
> brating the Sacrifice of the Mass; and that all excom-
> munications, all suspensions, all the punishments which
> a priest might undergo because he celebrated this Mass
> would be utterly null and void, *in futuro, in perpetuum* —
> in the future and forever.[14]

In 1986, Pope John Paul II designated a council of nine cardinals to study the issue of whether the old Mass had been abrogated. Eight of the nine concluded that it had not.[15] Refusing
to go along with the new Mass placed His Excellency in the
company of Saint Padre Pio who also refused to do so under
great pressure from his superior, although the version rejected
by Padre Pio was the far less radical 1965 version.

Archbishop Lefebvre clearly was not keeping with the spirit of
the times. For an archbishop of such dignity to refuse to go with
the flow was especially problematic because it made the other
bishops look irresponsible. Accusations flew, but despite these
accusations, or perhaps because of them, Lefebvre attracted a
worldwide following. The more his fellow bishops accused him,
the more the faithful turned to him.

Where was Pope John Paul II in all this? When he ascended
the Chair of Peter in 1978 he moved very slowly on the question

---

[14] Lefebvre, *Open Letter*, 148.

[15] The nine cardinals were: Ratzinger (the future Benedict XVI), Mayer, Oddi,
Stickler, Casaroli, Gantin, Innocenti, Palazzini, and Tomko. They answered the
following two questions: 1) Did Pope Paul VI authorize the bishops to forbid the
celebration of the traditional Mass? 2) Does the priest have the right to celebrate
the traditional Mass in public and in private without restriction even against
the will of his bishop? They unanimously agreed that Pope Paul VI never gave
the bishops the authority to forbid priests from celebrating the traditional rite
of Mass. In response to the second question, they stated that priests cannot be
obligated to celebrate the new rite of Mass; the bishops cannot forbid or place
restrictions on the celebration of the traditional rite of Mass, whether in public
or in private. For more details, see Peter Kwasniewski, "Minutes from the Commission of Cardinals That Advised John Paul II to Lift Restrictions on the Old
Missal," *New Liturgical Movement*, January 9, 2023, www.newliturgicalmovement.
org/2023/01/minutes-from-commission-of-cardinals.html.

of the faulty implementation of Vatican II. Radical, extreme alterations to the life of Catholics had caused so many to lose faith already that by 1980 the Holy Father himself felt the need to issue an apology:

> I would like to ask forgiveness in my own name and in the name of all of you, venerable and dear brothers in the episcopate, for everything which, for whatever reason, through whatever human weakness, impatience or negligence, and also through the at times partial, one-sided and erroneous applications of the directives of the Second Vatican Council, may have caused scandal and disturbance concerning the interpretation of the doctrine and the veneration due to this great Sacrament.[16]

So, he understood, yet he hesitated. One might say that his "eyes took some time to adjust," as he came out from behind the Iron Curtain. Regardless of these problems with the liturgy, everywhere he looked he saw "Springtime" for the Church. He had suffered through his entire adult lifetime, living as a suspect, operating in secret, running clandestine seminaries, having to hide from the government for church functions which we consider to be normal and routine, and now it was Springtime! He was free to speak, free to lead the Church to the worthy goal of salvation of souls. Everywhere he traveled he was met by large crowds, so he did not comprehend the collapse. Despite the distorted implementation of Vatican II, he did not share the gloomy perspective of millions of Catholics in the West who saw the postconciliar plummet and wondered what "Springtime" he was referring to.

Archbishop Lefebvre corresponded frequently with Cardinal Gagnon and Cardinal Ratzinger, the future Pope Benedict XVI, during the early years of the pontificate of Pope John Paul II. Lefebvre was never certain that his followers would be protected after his departure. He knew that his time on earth was limited, and he became concerned for the protection of the SSPX. Very often his priests were rejected by local ordinaries. He flew across the world for confirmations, to provide moral support, and to visit apostolates. He became increasingly concerned about what provision would be made for them. Without a successor to

---

[16] John Paul II, Apostolic Letter *Domenicae Cenae*, February 24, 1980; see Michael Davies, *Liturgical Shipwreck* (TAN Books, 1997), 21.

carry on his work, traditional Catholics would be forced into the destructive liturgical and doctrinal environment he had worked so hard to protect them from.

Over time, he found an ally in Bishop Antonio de Castro Mayer who had been the bishop of the Diocese of Campos, Brazil until his retirement in 1981. Bishop de Castro Mayer did not implement any of the "reforms" of Vatican II in his diocese, and after his 1981 retirement and replacement a parallel diocese completely dedicated to the practice of traditional Catholicism remained in place.[17] Bishop de Castro Mayer was helpful, but he was not young. For the SSPX to continue to exist after the death of Archbishop Lefebvre, auxiliary bishops would have to be appointed soon. Archbishop Lefebvre appealed repeatedly to the Holy Father, begging to be given auxiliary bishops, but his letters went unanswered. Finally, on July 8, 1987, he wrote to Cardinal Ratzinger in an emotional appeal to him to assist in securing auxiliary bishops:

> In order to prevent the auto-demolition of the Church we beg the Holy Father, through your mediation, to allow the free exercise of Tradition by procuring for Tradition the means to live and develop itself for the salvation of the Catholic Church and the salvation of souls: that the traditional foundations may be recognized, especially the seminaries; that His Excellency de Castro Mayer and myself may consecrate some auxiliaries of our choice in order to give to the Church the graces of Tradition, the only source of the renewal of the Church.
>
> Eminence, after almost 20 years of pressing requests that the "experiment of Tradition" be encouraged and blessed, requests always left unanswered, this is probably the final appeal in the sight of God and of the Church. The Holy Father and yourself will bear the responsibility of a definitive rupture with the past of the Church and its magisterium.[18]

Pope John Paul II had Cardinal Ratzinger write back to the aging archbishop, describing a proposal that would allow for the

---

[17] This "parallel diocese" eventually became the Personal Apostolic Administration of Saint John Mary Vianney. The history of this episode in the preservation of the traditional Latin Mass is covered extensively in Dr. David Allen White's *The Mouth of the Lion: Bishop Antonio De Castro Mayer and the Last Catholic Diocese* (Angelus Press, 1998).

[18] Laisney, *Archbishop Lefebvre and the Vatican*, 22, translation corrected.

continued use of the 1962 liturgy, and the right to "train seminarians ... according to the particular charism of the Society." However, the issue of auxiliary bishops could not be addressed until the tense relations with the Holy See were resolved. The letter also contained a stern warning—not to proceed with plans for securing auxiliary bishops without the approval of the pope:

> Excellency, do you find my words severe? I would have liked to express myself in another way, but the gravity of the matter at stake does not give me any other choice. Anyhow, I am sure you acknowledge the generosity of the proposal which is made to you in the name of the Holy Father, and which constitutes a real means to safeguard your work in the unity and catholicity of the Church.[19]

The letter, dated 28 July 1987, indicated that a Cardinal Visitor would be dispatched to visit the SSPX and find a suitable juridical status in conformity with canon law. Lefebvre wrote back in October, expressing optimism over the new development, and indicating, in a postscript, his desire that the Cardinal Visitor be Cardinal Gagnon. He also suggested possible locations for meeting the Cardinal Visitor:

> In order to go further towards a solution, it seems indispensable to meet with the Visitor, either by his coming to Ecône or Rickenbach, in Switzerland, or by our meeting him at Albano, in order to be able to study possible concrete means of this definitive solution.[20]

Cardinal Gagnon visited the Society November 11 at Ecône, staying for one month with his delegation. Monsignor Camille Perl visited schools, priories, and mother houses, then together Msgr. Perl and Cardinal Gagnon visited more schools, monasteries, and apostolates. At the end, Cardinal Gagnon announced:

> We have been struck everywhere by and keep a great admiration for the piety of the persons, for the relevance and importance of the works, especially with regards to catechesis, education, and the administration of the sacraments. We certainly have in hand all that is necessary to make a very positive report.[21]

---

19 Laisney, 26.
20 Laisney, 28.
21 Laisney, 39.

Two months later, an anxious Archbishop Lefebvre received a letter from Cardinal Gagnon dated February 15, 1988:

> Very Dear Monseigneur,
>
> After a long wait I was able to ask the Holy Father what had been done with regard to the Society of Saint Pius X and the wider problem of Tradition.
>
> He has confirmed that he had attentively read my long report and the propositions that you had given me.
>
> As usual, he had been very busy with problems of world-wide dimensions. But he has already requested some canonists to suggest juridical forms that could be applied to the Society. He should be able to present some projects for this and for the doctrinal problems before the end of April.
>
> He has asked me to give you this assurance and to invite you to patience. He would also like you to request your collaborators to have a great discretion in public declarations, indeed those who do not desire the reconciliation are happy to take advantage of the least thing to raise up opposition....[22]

The constant appeals for patience that various officials made to the aging archbishop led him to believe that he was being stalled. Given the hostility directed toward him from his fellow bishops, he saw danger in allowing himself to be stalled. He immediately wrote back to the Holy Father, thanking him for the visit, but going on to express his concern that "It would be regrettable if the hopes raised by this visit turned into disappointment, observing the continual delays in the application of even a temporary solution...."[23]

He suggested some key points for a successful solution, to include a Roman Secretariat composed of members chosen from within the SSPX, exemption from the local ordinaries, and consecration of several bishops by June 30:

> This second point is the most urgent one to be resolved, given my age and my fatigue. It is now two years that I have not done any ordinations at the seminary in the United States. The seminarians ardently aspire to be ordained, but I no longer have the health to be crossing

---

[22] Laisney, 41.
[23] Laisney, 42.

> oceans…. This is why I entreat Your Holiness to resolve
> this point before June 30 of this year.[24]

The urgency of the situation was clear. In 1987 the archbishop gave 2,500 confirmations in France alone. In one 1984 ceremony in Chile, there were 1,527 *confirmandi*. The SSPX now had 530 places of worship on five continents. And the archbishop was 83 years old.

A meeting was arranged for mid-April 1988, the result of which was a solution that all participants could agree upon. However, it provided for only one bishop, which, given the size of the SSPX and the number of faithful, was barely adequate. Archbishop Lefebvre wrote to Cardinal Ratzinger: "The prospect of having a successor in the episcopate gives me great joy and I thank the Holy Father and yourself for it. Only one bishop will hardly suffice for the heavy workload; wouldn't it be possible to have two, or at the least, couldn't the possibility of raising its number in the next six months or a year be provided for?"[25]

The response from Cardinal Ratzinger, dated April 28, 1988, must have been more than the archbishop could take. A few key lines clearly illustrate what was happening:

> …Now this requires common study and reflection and
> could take still more time…
>
> …Thus a definitive answer cannot be given to you
> for the moment, but it will be at latest in the first half
> of June…
>
> …With regard to nomination of a bishop, the Holy
> Father tends to regard your proposition taking into
> account the practical and psychological reasons for
> such a nomination. However, this one could not happen
> right now, even if there were no other reason than the
> preparation and examination of the files according to
> the usual procedure of episcopal nominations.
>
> …though the definitive solution must wait some
> while because such an important problem cannot be
> resolved by being treated with precipitation…[26]

A protocol was reached, and on May 5 the archbishop signed it. However, it was most restrictive, and failed to adequately

---

24 Laisney, 43.
25 Laisney, 65.
26 Laisney, 68.

address his concerns in two areas: it was vague about the date of an eventual episcopal consecration, and it would leave this bishop powerless, since all jurisdiction would come from the local ordinary.

It soon became clear, at least to Archbishop Lefebvre, that there was no real intention of ever providing a bishop for the SSPX when Cardinal Ratzinger's secretary, Fr. Klemens, gave the archbishop a draft letter to sign. The letter, intended to be from Archbishop Lefebvre to the Holy Father, was full of apologies, pleas for forgiveness for "my behavior and that of the Society," to all of which he was willing to sign his name, but it contained the following deadly language:

> Lastly, I wish to express my gratitude for the intention that you manifested to take into account the particular situation of the Society, proposing to nominate a bishop chosen from its members, and especially in charge of providing for its specific needs. *Of course, I leave to Your Holiness the decision concerning the person to be chosen and the opportune moment. May I just express the wish that this be not in the too distant future?*[27]

After a sleepless night, he wrote to Cardinal Ratzinger. It seemed that in good conscience he could not postpone the episcopal consecrations again. He now believed that he was being dragged along, delayed, until it would be too late for him to do anything. This would be the fourth time the event was being postponed. The date of June 30 had long ago been set, and it was in fact the latest possible. A file of candidates had already been provided; there were still two months within which the Holy Father could issue the mandate.

The rest is history. Communications broke down. Instead of responding to the aging archbishop's request with a sense of urgency, Cardinal Ratzinger and the Holy Father demanded obedience. They simply did not see any such "crisis" or state of emergency that the archbishop seemed to see so clearly; they did not see the need for one bishop, much less several of them. And the archbishop, convinced at this point that a crisis existed which was not fully understood by the Holy Father, trusted that the penalty of excommunication, which he knew he would

---

[27] Laisney, 81, emphasis added.

automatically (*latae sententiae*) incur, would be held invalid by Him Who Judges All.

On June 30, 1988 he consecrated four candidates of his choice at Ecône; they signed the Anti-Modernist Oath that was once required of all bishops, and Bishop de Castro Mayer assisted at the ceremony. A hush settled over Rome that day, and the mood was described as similar to when Rome receives the news of the death of a pope. Pope John Paul II tried to prevent the consecration, acting far too late and in a way that was awkward and unfortunate. His gesture was described at the beginning of Lefebvre's consecration sermon:

> Yesterday evening, a visitor came, sent from the Nunciature in Berne, with an envelope containing an appeal from our Holy Father the Pope, who was putting at my disposal a car which was supposed to take me to Rome yesterday evening so that I would not be able to perform these consecrations today. I was told neither for what reason, nor where I had to go! I leave you to judge for yourselves the timeliness and wisdom of such a request.
>
> I went to Rome for many, many days during the past year, even for weeks; the Holy Father did not invite me to come and see him. I would certainly have been glad to see him if some agreement would have been finalized....[28]

Shortly thereafter the Holy Father issued *Ecclesia Dei Adflicta*, in which he asked "bishops and all those engaged in the pastoral ministry of the Church" to be generous in allowing the 1962 Missal, and to respect the "rightful aspirations" of those requesting it. In most cases the document was ignored by bishops. As for the requests from the laity, we see how they were treated.

Soon small groups of priests who had been with Archbishop Lefebvre began to organize and approached the Holy Father with their desire to establish societies that would work with bishops, those few who were willing, to provide the sacraments to Catholics according to the old rite. One of them, the *Fraternitas Sacerdotalis Sancti Petri* (FSSP), would grow quickly over the next few years. The Institute of Christ the King Sovereign Priest was canonically erected on September 1, 1990 by Monsignor Gilles Wach in Gabon, Africa, where Archbishop Lefebvre had carried out missionary work. The archbishop was still remembered

---

[28] Laisney, 117.

fondly enough that a few years later, in 1996, a postage stamp of his image was issued there. Both the SSPX and the Institute of Christ the King have missions in Gabon today.

The status of the old Mass in 1988, after this critical juncture, was that it had not been abrogated, and that any priest had the right to say it, but the Holy Father would not state this publicly. He simply asked bishops to be generous with it, and his good nature inclined him to count on their good will. For the time being, there would be no clarification of this simple fact: the old Mass had never been forbidden. The myth was allowed to propagate that laymen incurred the penalty of excommunication for assisting at Masses offered by the SSPX, when in fact no penalty was incurred. A lay person had every right to assist at the Mass, especially in those situations in which the local ordinary (bishop) did not provide an "indult" Latin Mass, which most did not. This simple fact would not be publicly known until 2002 when it was finally admitted by the *Ecclesia Dei* Commission.[29]

Thus, in 1995, I still mistakenly believed that we were forbidden to attend Mass with the SSPX and that, furthermore, to attend Mass in the SSPX chapels was an act of schism. This misunderstanding laid the groundwork for what was to follow.

---

[29] Letter from Msgr. Perl, *Ecclesia Dei* Commission, September 27, 2002. For a fuller analysis of this matter, see Peter Kwasniewski, "Is It Ever Okay to Take Shelter in an SSPX Mass?," *OnePeterFive*, April 3, 2019, https://onepeterfive.com/sspx-mass-shelter/; Paul Casey, "SSPX Masses and Fulfilling the Sunday Obligation," *Catholic Family News*, April 28, 2023, https://catholicfamilynews.com/blog/2023/04/28/sspx-masses-and-fulfilling-the-sunday-obligation/.

CHAPTER 10

THE GENERAL'S EXECUTIVE OFFICER

SHORTLY AFTER MY EXCHANGE WITH FR. KAISING, I received the long-awaited reply from the Vatican. My normally stable hands trembled as I opened the letter. It was from Cardinal Felici's assistant, Msgr. Camille Perl. I had never received a letter from the Vatican before! The letter was dated April 23, 1997. After apologizing for the lengthy delay, Msgr. Perl said:

> We note that you have respectfully and consistently explored every possible avenue in order to secure permission for the celebration of the traditional Latin Mass at Fort Bragg and that this permission has been consistently denied. If the commission were to address anyone about your request, it would be Archbishop Dimino of the Military Services. It is not the practice of this Pontifical Commission to write to a military superior, even if he happens to be Catholic.
>
> Further, we do not judge that a letter to Archbishop Dimino would serve any useful purpose at this time. We say this particularly in light of the fact that those petitioning the celebration of the traditional Latin Mass number only twenty-eight persons which include five members of your family and seven members of the Donnelly family.[1] If, at some time in the future, a *significantly larger* number of the faithful should manifest a desire for such a celebration, it might be appropriate to renew your petition.[2]

It seemed that there was always some very good reason that our request could not be satisfied. If no priest was available, that would serve as a convenient excuse. If a priest was available

---

[1] Not their real name.
[2] Emphasis added.

PONTIFICIA COMMISSIO
« ECCLESIA DEI »

N. 9/95

Rome, 23 April 1997

Maj. David L. Sonnier
105 Frazier Court
Fort Bragg, North Carolina 28307
U.S.A.

Dear Major Sonnier,

Thank you for your letter of 27 February 1997 addressed to His Eminence Cardinal Felici. It has only recently arrived in the offices of this Pontifical Commission. We do not know the specific reason for the postal delay, but are aware that since the beginning of this year mail from the United States has sometimes arrived only after considerable delay.

Your most recent letter and the enclosed material which you have submitted have been the object of attentive study by this Pontifical Commission. On the basis of that data, we would like to make the following comments.

1. We are genuinely sympathetic to your desire and that of those whom you represent for the celebration of the Mass according to the 1962 Roman Missal. It must be stated clearly, however, that this Pontifical Commission has not been granted among its faculties the right to impose decisions on diocesan Bishops. The *Motu Proprio* explicitly states that the Commission is "to collaborate with the bishops, with the Departments of the Roman Curia and with the circles concerned, for the purpose of facilitating full ecclesial communion of priests, seminarians, religious communities or individuals until now linked in various ways to the Fraternity founded by Mons. Lefebvre" (*Ecclesia Dei* 6, a).

2. We note that you have respectfully and consistently explored every possible avenue in order to secure permission for the celebration of the traditional Latin Mass at Fort Bragg and that this permission has been consistently denied. If the Commission were to address anyone about your request, it would be Archbishop Dimino of the Military Services. It is not the practice of this Pontifical Commission to write to a military superior, even if he happens to be Catholic.

3. Further, we do not judge that a letter to Archbishop Dimino would serve any useful purpose at this time. We say this particularly in light of the fact that those petitioning the celebration of the traditional Latin Mass number only twenty-eight persons which include five members of your family and seven members of the ████████████ family. If, at some time in the future, a significantly larger number of the faithful should manifest a desire for such a celebration, it might be appropriate to renew your petition.

Wishing you God's blessings during this joyful Easter season, I remain

Sincerely yours in Christ,

Msgr. Camille Perl
Secretary

(there were several by now), then the fact that the local ordinary was opposed to the Latin Mass could be employed against us. If not that, there were too few signatures. Had we sent him 1,000 signatures, what would the excuse have been?

As a side note, in 1997 the number of professed pagans and Wiccans in the military was miniscule, but by a decade later the number came out to around one tenth to three tenths of *one percent* of the 1.5 million servicemembers throughout the world. There might be five such worshippers at any one time at a base as large as Fort Bragg or Fort Hood. Nevertheless, Fr. John Kaising, a priest in the Roman Catholic Church, a Colonel in the US Army, and the executive officer of the US Army Chief of Chaplains, had not only gone to bat for them so that they could worship according to their needs, he had also made a media event of it. Now we had a letter from the *Ecclesia Dei* Commission in the Vatican stating that they couldn't support our request for a Latin Mass at Fort Bragg because there were still not enough of us to justify it.

I was the commander of a company when I began collecting signatures and now I was the executive officer of a battalion of several hundred soldiers. I could have easily used my influence to get many more signatures, but instead I actively sought out people who genuinely desired the presence of the Latin Mass. And I didn't petition while on the job.

I decided that I would try to continue discretely collecting more signatures, since the insufficient number was the latest excuse. Let them continue to make excuses; I would come after them one at a time until we prevailed. But first, perhaps it would be best to wait for Fr. Marceaux to leave; he would be gone soon, and things would certainly be better.

In looking for ways to deal with this bizarre situation, I discovered that, in addition to an SSPX chapel that was a bit too far away for us, there were "independent Catholic chapels" in the local area. At this point I could see no reason for not occasionally attending Mass at one of these locations. This would let us experience the liturgy we had come to love, and which we were rarely able to find. After attending a few times, we discovered that the priest, who shall remain unnamed, was agreeable to coming to Fort Bragg once a month to offer Mass in our quarters. I wrote to Michael Davies to get his opinion on the matter. His response is reproduced here, leaving out the name of the priest who eventually began offering an occasional clandestine Mass in the old rite for us:

31 May 1997

Dear Major Sonnier,

I have just returned from the Chartres Pilgrimage and subsequent Remnant Pilgrimage to Catholic shrines in France. Chartres was a great success, about 16,000 pilgrims present with an average age of twenty, the Mass celebrated by Cardinal Felici with the Papal Nuncio and Bishop of Chartres present, and a message and blessing from the Holy Father.

I had hoped to answer your letter before leaving for Chartres, but I had some articles and a book to complete, and I could not manage it.

The *Ecclesia Dei* attitude to the number of signatures did not surprise me as the Commission has done this on previous occasions. I agree completely with what you say about your position ruling out an aggressive signature collecting campaign.

As regards Father (X), I think that you are perfectly entitled to have recourse to him. It is the manifest will of the Holy Father that those wishing to have recourse to the 1962 Missal should have the traditional Mass available to them. Your chaplain is defying the Pope and depriving you of what is yours by right, and in providing the Mass for you, Father (X) would be acting in accordance with the Pope's wishes.

He went on to suggest that I should not be the one to extend the invitation to Father X, but this suggestion was out of concern for my career and potentially losing my retirement. I was not concerned about that. The letter continued:

His Mass would definitely fulfill the Sunday obligation as the New Code of Canon Law states that the obligation is fulfilled by being present at Mass in any Catholic rite. Under the Old Code such a Mass would not have fulfilled the obligation as it listed in meticulous details the places where the obligation could be fulfilled.

Father (X), like every priest of the Roman Rite, is entitled to use the 1962 Missal, and if, as is possible, *Quo Primum* no longer applies, he has a right to celebrate the

24 Cromwell Avenue, BROMLEY, Kent, BR2 9AQ, England

'phone 44 181 402 2248   Fax 44 181 289 6377

31 May 1997

Dear Major Sonnier,

I have just returned from the Chartres Pilgrimage and subsequent Remnant Pilgrimage to Catholic shrines in France. Chartres was a great success, about 16,000 pilgrims present with an average age of twenty, the Mass celebrated by Cardinal Felici with the Papal Nuncio and Bishop of Chartres present, and a message and blessing from the Holy Father.

I had hoped to answer your letter before leaving for Chartres, but I had some articles and a book to complete and I could not manage it - at least I think that I did not. If I have already replied put it down to premature senility.

The ED attitude to the number of signatures did not surprise me as the Commission has done this on previous occasions. I agree completely with what you say about your position ruling out an aggressive signature collecting campaign.

As regards Father ████████, I think that you are perfectly entitled to have recourse to him. It is the manifest will of the Holy Father that those wishing to have recourse to the 1962 Missal should have the Traditional Mass made available to them. Your chaplain is defying the Pope and depriving you of what is yours by right, and in providing the Mass for you, Father ████████ would be acting in accordance with the Pope's wishes. I would suggest, however, that the invitation be issued by someone other than yourself and that you should not personally enter into any correspondence with him. If asked by your chaplain, you could then answer truthfully that Father ████████ was not invited to the base by you. His Mass would definitely fulfil the Sunday obvligation as the New Code of Canon Law states that the obligation is fulfilled by being present at Mass in any Catholic rite. Under the Old Code such a Mass would not have fulfilled the obligation as it listed in meticulous details the places where the obligation could be fulfilled.

Father ████████, like every priest of the Roman Rite, is entitled to use the 1962 Missal, and if, as is possible, *Quo Primum* no longer applies, he has a right to celebrate the traditional Mass derived from immemorial custom. A few years ago in *The Latin Mass* I wrote an article dealing with this point jointly with Count Capponi.

In your position I would under no circumstances leave the Army. You should remain in the service until you qualify for a pension unless, of course, one of the civilian jobs you mentioned would pay tremendously more. It would also

1

concede victory to your very uncatholic chaplain if you retired. If you keep fighting and win the day it would establish a precedent and could result in the Traditional Mass being celebrated in other Military Bases.

I am writing this letter in my personal capacity and not as Una Voce President. Strictly-speaking, UV is concerned only with obtaining Masses with episcopal permission, and if it became known that I had recommended assisting at the Mass of a priest who does not have such approval it could harm the Federation.

Please do not hesitate to contact me by letter, fax, or 'phone if ever you feel that I can be of help. I shall certainly raise the Fort Bragg case the next time that I am in Rome.

With every good wish *in Domino*,

Michael Davies

traditional Mass derived from immemorial custom. A few years ago in *The Latin Mass* I wrote an article dealing with this point jointly with Count Capponi.

In your position I would under no circumstances leave the Army. You should remain in the service until you qualify for a pension unless, of course, one of the civilian jobs you mentioned would pay tremendously more. It would also concede victory to your very uncatholic chaplain if you retired. If you keep fighting and win the day it would establish a precedent and could result in the traditional Mass being celebrated in other Military Bases.

I am writing this letter in my personal capacity and not as Una Voce President. Strictly speaking, UV is concerned only with obtaining Masses with episcopal permission, and if it became known that I had recommended assisting at the Mass of a priest who does not have such approval it could harm the Federation.

Please do not hesitate to contact me by letter, fax, or 'phone if ever you feel that I can be of help. I shall certainly raise the Fort Bragg case the next time that I am in Rome.

With every good wish *in Domino*,

Michael Davies

For the remainder of our time at Fort Bragg we would periodically have this priest, Father X, offer Mass in our residence on Frazier Court in the Field Grade Officer Housing Area at Fort Bragg. Others joined us, but it was always a small and discreet group. The priest usually stayed for dinner with my family. Initially one of the young officers from New York served the Mass, but eventually I taught my oldest son the responses, and on some occasions the two of us served the Mass together. The Mass was not authorized by the diocese, nor was it authorized by the Archdiocese for Military Services, but I was no longer under the impression that they were dealing with us in good faith. To this date I am convinced that establishing a private chapel for the old rite, such as we did on this occasion, is the best way to preserve our faith when surrounded with clergy who are using the modern liturgy to undermine it.

Fr. Marceaux departed for an assignment in Belgium soon thereafter, and we were glad to see him go. His replacement, Fr.

Frank Whalen, arrived sometime in the spring of 1997. Hoping to get off to a good start with him for a better result, I contacted his office with an informal request and an offer to discuss the request with him. The answer that came back, through a mediator, was "absolutely not." It seemed that he was as hostile as his predecessor, who had probably told him that he could expect to hear from me. The reason he cited for denial of the request was that the bishop of Raleigh did not allow the Tridentine Mass so he could not either. Since this was his logic, I called the office of the bishop of Raleigh to see whether he really cared one way or another about us having a Latin Mass on Fort Bragg. In fact, he didn't care one bit:

"I don't care what you do on the military base," he said.

"Would you allow a Latin Mass in the Diocese of Raleigh?" I asked.

"I won't allow it in my diocese, but Fort Bragg is not in my diocese."

Armed with these words from Bishop Gossman, I prepared a petition and a cover letter to send to Fr. Frank Whalen. But soon I began to see the hopelessness of dealing with him. We quickly noted that he was not being careful, as Fr. Marceaux had been, to make his actions appear correct and in accordance with Church doctrine or canon law. In fact he was quite reckless. He went out of his way to disrupt the daily Mass crowd. My wife had been taking our three children to daily Mass once a week as part of the homeschool routine. It was easier to go to the Novus Ordo Mass on base than to the Maronite Church, a 25-minute drive away, and during the week there were no major liturgical tomfooleries by the retired priest who offered Mass. He was opposed to the Latin Mass for unknown reasons (probably because he knew he would not be welcome if he didn't publicly oppose it) but he seemed orthodox.

Apparently, this little bit of orthodoxy was too much for Fr. Frank Whalen to tolerate. Upon his arrival, Fr. Whalen took over the daily Mass to better ensure that those faithful old war widows, soldiers' wives, homeschooling mothers and their families, and one-legged veterans wouldn't be coming for much longer to seek comfort in his chapel. "This is *not* the Church of Archbishop Fulton Sheen! This is the church of the '90s!" he announced on one occasion. "This is not the Church of the '50s, this is the church of the '90s!" he proclaimed on another occasion.

If someone from the congregation didn't get up to do the epistle reading, which nobody who attended daily Mass wanted to do, he just sat and waited. For ten minutes, if necessary, he would just sit there waiting for someone to get up and go read the epistle, finally barking that "those old days are over! They're over! It's your responsibility now!"

One evening Fr. Whalen was speaking at the chaplain center on a variety of issues with any Catholics who wanted to listen. Working late, I walked in to retrieve my family just in time to witness the most unbelievable sight. My dear wife, Lorri, stood near the back of the conference room holding one child, while one slept on the floor, and another played nearby. There were about 150 people listening as she debated the senior Chaplain. This bizarre confrontation was taking place in front of stunned military Catholics of all ranks. The debate began over the ordination of women, which he was claiming would happen soon "for the good of the Church." A young Catholic friend standing nearby witnessed the whole debate. Since I walked in when it was half over, I asked him to write a summary to be forwarded to the Military Archdiocese. The letter, exactly as it was written by this courageous young soldier, reads:

> 2 April 1998
>
> After the events of last night's Catholic discussion at Fort Bragg, led by Chaplain Whalen, I decided once again to put pen to paper.
>
> At 7 p.m. Chaplain Whalen and Chaplain Honor opened it up, after a prayer, for general Q&A. A lady immediately asked, "What do you think is the prospect of women being ordained to the priesthood, and what's in store for the future . . . what is needed to allow women to be ordained?"
>
> Chaplain Whalen answered very straight-forward: "A dead Pope!"
>
> The audience was stunned. Chaplain Whalen gave the exact same answer [again]. Mrs. Sonnier immediately protested, saying that this Pope (as others before him) has stated that the Church has no authority to ordain women to the priesthood, and that priests represent Christ as the groom to the bride, the Church, and that women "priests" are an oxymoron. Chaplain Whalen said that the Church can change.

There is much more that I could tell of Fr. Whalen's behavior, but it would be beside the point.

In the meantime, I had been selected to be the executive officer for the Commanding General of the U.S. Army Special Operations Command, Lieutenant General Peter J. Schoomaker. I was up-front with him during my interview; he asked me a

With my family about the time I began working as the Executive Officer for Lieutenant General Schoomaker, Spring 1997.

standard question—whether there was something about me, personally, that he needed to know, and I told him that I was engaged in a battle with dissident clergy who were trying to separate Catholics from the Church of Rome. I told him that we were interested in restoring the Latin Mass, the old liturgy that had existed before the Second Vatican Council, because some dissidents had taken advantage of the new liturgy for the purpose of spreading doctrinal error. I was quite blunt about it. He offered me the position anyway.

Lt. Gen. Keane and Lt. Gen. Schoomaker held equal rank but had very different responsibilities. Keane was the commander of the XVIII Airborne Corps, which included the 82nd Airborne Division at Fort Bragg, the 101st Division at Fort Campbell, Kentucky, and the 10th Mountain Division at Fort Drum, New York. Schoomaker commanded the Special Forces groups and all other special operations units scattered, quite literally, around the world. I was overwhelmed by the scope of the job as his executive officer, but I'd been around long enough by now to know that I would figure out how to do it.

The first few weeks were a blur of activity. As the executive officer, I had to prioritize the general's workload while he was in garrison, handle his correspondence, make his travel arrangements, and keep his schedule. He traveled frequently, and when he did, it was normally the aide-de-camp that traveled with him. But the aide was experiencing some health problems within his family that made it difficult for him to travel. So, it wasn't long before I ended up making some of the trips with Lt. Gen. Schoomaker. He was a decent and kind man, although he could be tough to get along with if one fouled up any paperwork or made arrangements that weren't completely in accordance with his guidance.

Despite his kindness and excellent mentorship, and despite the obvious successes I was having as a career officer, the spiritual battle caused me to question why I served in the Army. This problem with our chaplains, which should have been so easily resolved by simply appealing to Church authority, was only getting worse. I finally put a written request through to Fr. Whalen to get an answer on paper as to what his rationale was for not allowing us to have a Latin Mass. His response follows:

DEPARTMENT OF THE ARMY
HEADQUARTERS, XVIII AIRBORNE CORPS AND FORT BRAGG
FORT BRAGG, NORTH CAROLINA 28307-5000

August 1, 1997

REPLY TO
ATTENTION OF:

AFZA-CH

Major and Mrs. David L. Sonnier
105 Frazier Court
Fort Bragg NC 28307

Major and Mrs. David L. Sonnier,

I want to answer your primary question first. I am not interested in having a Latin Mass celebrated at Ft. Bragg.

Secondly, I am well aware that Bishop Gossman has no jurisdiction over Ft. Bragg. I have been an Army Chaplain for twenty-five years, and understand the relationship between the installation and the local diocese.

This is my fourth assignment at Ft. Bragg since 1982. During that time our Catholic Community has always had a very good and friendly relationship with Raleigh Diocese. We have adopted many of their policies because they are the right thing to do and we avoid contradictions in the mind of our military and civilian communities. We have also used many services they provide which have been a tremendous spiritual benefit to countless soldiers and families. I am not willing to jeopardize this very good relationship. It would be too great of a loss for too many people.

I have discussed the issue of a Latin Mass at Ft. Bragg with my brother priests now assigned here. We all agree with Raleigh Diocese's policy.

In Christ,

Frank J. Whalen
Chaplain (COL) USA
Corps and Installation Chaplain

August 1, 1997

Major and Mrs. David L. Sonnier,

I want to answer your primary question first. I am not interested in having a Latin Mass celebrated at Fort Bragg.

Secondly, I am well aware that Bishop Gossman has no jurisdiction over Ft. Bragg. I have been an Army Chaplain for twenty-five years, and understand the relationship between the installation and the local diocese.

This is my fourth assignment at Ft. Bragg since 1982. During that time our Catholic Community has always had a very good and friendly relationship with Raleigh Diocese. We have adopted many of their policies because they are the right thing to do and we avoid contradictions in the mind of our military and civilian communities. We have also used many services they provide which have been a tremendous spiritual benefit to countless soldiers and families. I am not willing to jeopardize this very good relationship. It would be too great of a loss for too many people.

I have discussed the issue of a Latin Mass at Ft. Bragg with my brother priests now assigned here. We all agree with Raleigh Diocese's policy.

In Christ,

Frank J. Whalen
Chaplain (Col.)

I decided I'd go visit Fr. Whalen and discuss this personally. Even though he didn't seem to want to discuss the issue, I would confront him with kindness and logic. I would present my argument by expressing my love and appreciation for the old Mass. It would be easy to show him what the Holy Father had said about it, to explain about the priests living nearby who could offer it, and to tell him about the seminaries full of young men who were being ordained to offer Mass in the old rite.

From the moment I entered his office I could tell that I would get nowhere, but I felt obligated to try anyway. He jumped up from behind his desk, crossed his arms, and veins jumped out in his forehead, from which sweat began oozing. There was a wild look in his eyes, like someone had just stepped on his foot.

"What do you want!" he shrieked.

"Sir, I came to talk about our request."

"There's nothing to talk about!"

"But there's plenty; I have some information … if you'd like I can leave it with you…"

"Anything you give me is going right in the trash can. I won't even look at it."

"But what you said about the shortage of priests—it's not true. There are orders of priests that follow the old rite, and their seminaries are full. I can give you the information right here."

"I'm not interested in that. That's not the direction the Church is going. Those days are over!" he bellowed.

"But we need more priests in the military, if you could just accommodate…"

"God is not calling men to the priesthood anymore! He used to, but no more!"

"But He is. And the pope, the Holy Father, has asked you to be generous...."

"He hasn't asked me anything! I answer to one man, and one man alone!"

By now, spit was flying as he shrieked. He wiped his profusely sweating brow.

"But Cardinal Ratzinger..."

"I don't answer to him! Go to him!" he screamed. "Don't come to me!"

I was stunned. I waited for some inspiration; some words perhaps sent from God that would help me to overcome this impasse. I slowly prayed an Ave Maria while waiting for him to gather his wits. He sat down behind his desk and pretended to be returning to a report. After a few seconds he looked up:

"This meeting is over. Finished! Get out of here!"

I didn't leave. He stood again, face reddening.

"Get out of here!" he yelled. "Get out of my office! You piss me off!"

I still didn't leave.

"You are dismissed, Major! This meeting is over! Now get out! Go!"

I stayed.

He reached for the phone.

"Listen, Major, you get the hell out of my office now or I'm calling the military police!"

I left. The best thing I could do at this point was to preserve all meeting notes and correspondence for future reference.

CHAPTER 11

DE OPPRESSO LIBER

ON THE WAY BACK OVER TO USASOC HEAD-quarters I did some serious soul-searching. What was I doing serving in this Army? A few years ago I would have happily given my life for the noble cause of defending a great nation; a good and decent people. And now...?

The fearlessness and raw courage I had as a young officer was gone. Why should I lay down my life in the service of an army that wouldn't permit me a legitimate option of religious worship? And what kind of priests (and soldiers) were these chaplains who were so infuriated by our simple requests? How could I go into battle trusting my last confession to a dissident priest? Could I ever confide in one of them at my moment of death? If they had so little regard for what the Church teaches regarding liturgy, how could I trust in their knowledge of death, judgment, heaven, and hell? If we couldn't even agree with our priests on the fundamentals of the Catholic faith, how could we ever work together to accomplish something against a common enemy? Clearly the chaplain in the office that I just left saw me as his real enemy.

These were not like the chaplains who had been on the beaches of Normandy comforting dying soldiers or flew through dangerous zones to bring the Eucharist to isolated base camps in Vietnam. They lacked that kind of courage. It was easier to bully a lowly Major trying to remain loyal to the Church than to confront the multitude of dissidents clamoring for women priests, insisting on girl altar boys, and pressing for radical changes in liturgy and doctrine. My simple request reminded them that they had become lazy about saving souls in a country that was turning its back on God. Rather than fighting the inertia to live up to their office, possibly facing ridicule from liberals and modernists, they chose the easy way out.

Years ago, a priest serving in the Persian Gulf War (1991) had offered Mass in the old rite. This priest served as chaplain of a medical support unit in a hospital with the First Marine Division. He had loved the old Mass from his seminary days. During a liturgy class at Mt. Saint Mary's Seminary in Emmitsburg, Maryland, he wrote a paper about offertory prayers. It was then that he saw the offertory prayers from the preconciliar rite for the first time. The conclusion he reached in his paper caused his professor great concern, and while he received a good grade for it he also received a two-page commentary on how "dangerous" his thinking was. He continued to study the old Mass, bringing a *St. Andrew's Missal* to Mass and saying the old prayers during the new Mass, until his seminary professors forbid it. After his ordination in 1985, he continued to study the old Mass, entered the reserves, and was called up to active duty in 1990. On some occasions he would say a Latin Mass for a small group of Navy and Marine personnel (on one occasion, at 4:00 in the morning in a tent). After the war he went back to the reserves and continued to drill one weekend a month. Eventually he joined the FSSP. When that happened, Archbishop Dimino, of the Military Archdiocese, let him keep his commission, but removed him from the reserves. Why did Archbishop Dimino do this? Was he concerned that if he allowed a single priest to pray in Latin it would expose the dishonest narrative that had been propagated since the Council? How many good priests were prevented from serving as military chaplains, even though there was a terrible shortage?

By now, I had a good enough relationship with Lt. Gen. Schoomaker that I could talk with him about this:

"You know, this has all happened before — in England," he said.

How was it possible that this general, who was a Protestant, could see so clearly what many Catholics were unable to see? He was quite right. The way the American Catholics were going, most of them were separating themselves from the Church of Rome through indifference. Dissidents had control of the media, control of many of the chancery offices, control of all the mechanisms for driving a wedge between the Catholic Church and the American Church — "AmChurch," as we had begun to call it. The only difference was that England had a monarch who turned his back on the Church. The U.S. is governed by "We the People," with most "Catholics" in this governing body having turned their back on the Church. The effect was the same.

"You should talk with Jack Keane about this," he said.

"Sir, I sent him a letter last year. He sent back a response full of mistakes."

"You mean, like typos?"

"No, sir, like he was dead wrong on several points."

Schoomaker became suddenly defensive. General officers have to stay on good terms with each other. A public disagreement between a couple of three-star generals at Fort Bragg would be a public relations disaster.

"Let me see his response."

I pulled it from my increasingly thick file of correspondence. He read it for a moment.

"What's wrong with this?"

"Sir, it says that Colonel Marceaux used the 1970 Missal 'which suppressed the traditional Latin Mass' as the basis for his judgment. It doesn't acknowledge the existence of more recent guidelines—such as the 1984 and 1988 guidance in which the pope asked bishops and clergy to be generous in allowing the Latin Mass. Also, he's telling us to drive to Virginia or Atlanta to attend a Latin Mass."

"Where? I don't see that."

"He says, 'I understand that you have a schedule of traditional Latin Mass celebrations in the United States, and I encourage you to pursue these as a means of meeting your personal religious needs.' But since the bishop of Raleigh has not complied with *Ecclesia Dei*, that means either Virginia or Atlanta are the nearest locations in which a bishop has allowed the Latin Mass in accordance with the pope's instructions. I can't drive four to six hours away for a Sunday Mass."

I told him about the existence of Society of Saint Pius X chapels in Raleigh and Goldsboro, which we had been told to completely avoid. I didn't tell him about the independent traditional Catholic chapels nearby, or the fact that there was occasionally a clandestine Mass in my residence a few hundred yards away.

"Sir, I gave General Keane the information he needed to make a decision. He listened to a dissident chaplain instead."

"Jack Keane is a good man—I don't think he would have done that intentionally. You need to go talk with him."

I made an appointment to see him, and I even prepared some briefing slides. He was cordial, offered me some coffee, and I sat across a small table from him and flipped through the briefing slides. He listened patiently as I showed him the key text from *Ecclesia Dei*, read quotations from Vatican II, presented statistics about the growth of Latin Mass communities around the country and the growth of the FSSP seminary, and summarized the history of the requests I had made to the XVIII Airborne Corps chaplains. I listed all the objections they had raised, in the order in which they had raised them, and showed why it was that each of them was wrong:

"The Latin Mass is just for the elderly and infirm."

On the contrary, it is for anyone who desires it. No such age limitation is ever mentioned in any Vatican document.

"There is not a priest available to offer a Latin Mass."

No, there are several.

"It would be divisive."

It would bring many disillusioned Catholics back into the Church. I would probably not be a practicing Catholic at this moment were it not for the restoration of the Latin Mass to a certain Church in Ohio.

"The bishop of Raleigh doesn't allow it."

I spoke with him myself, and he doesn't care what we do. I'll call him now if you'd like to speak with him.

The general looked as if he were about to go to sleep. Was he following the shifting logic of the denials?

"Do you remember the old Mass, sir?"

It took a few seconds for him to respond. He had the thousand-mile stare. Finally, he stirred and answered without looking at me. "Remember it well," he mumbled, as if talking to the floor.

"Do you go to Mass now?"

"No."

"Sir, would you be willing to help us?"

"Why is it that you don't just go to one of those other places on your list to attend Mass?"

"The nearest chapel is four hours' drive away, sir. All three bishops in the Carolinas are refusing to allow the old Mass. They've all received plenty of requests, and they're all refusing. North Carolina and South Carolina alike."

"Pete Schoomaker told me something disturbing—that you're considering declining command so that you can request a transfer away from Fort Bragg."

"Yes, sir."

"Why?"

"I don't believe I should have to drive four hours away for something the pope told bishops and clergy everywhere to generously allow. If the bishop of Raleigh is in rebellion from the pope's guidelines, then that's one thing. But for the Military Archdiocese to emulate his misguided example and force his policy on soldiers is wrong. For Catholic officers to allow it to happen is wrong. Our troops deserve better than this, and I can't live with it. I can't be a part of it. Anyway, I'm just doing what you and Chaplain Whalen told me to do. You said to go to one of the places on the *Ecclesia Dei* list, and I'd have to move to do that. So, I'll move."

"I said that?"

"Yes, sir."

There was a long, very hopeful pause. Then…

"The Catholic Church needs to change. Why is it that after all these years they are still, still, to this day, not ordaining women?"

"The Catholic Church doesn't have the authority to do that, sir."

He talked and talked. He talked about himself, his job, he talked about Fordham University, and a variety of irrelevant things. Then suddenly, the time he had allocated for this office call expired. Urging me to "hang in there," he escorted me out of the office.

LTG Schoomaker was signing a stack of awards when I returned to the office. "What did Jack say? Is he going to help you out?"

"I don't think so, sir."

There was one last hope. Archbishop Dimino, who never answered a single letter from me, went into retirement and suddenly we had a new archbishop of Military Services, Edwin F. O'Brien. Perhaps he would help. My wife and I both wrote appeals to him, asking him to assist us. My letter would be repetitive, but the letter written by my wife is worth reproducing:

> 18 August 1997
>
> Dear Archbishop O'Brien,
>
> I'm sure you are aware of our petitions to have a Latin Mass at Fort Bragg. Enclosed please find the latest answer we received from Fr. Whalen. Each time we have asked previously we have been given different reasons for denial which we have resolved. With this latest refusal, the first sentence states that Fr. Whalen is not interested in having a Latin Mass here at Ft. Bragg. Since the request must come from the head chaplain, is this

the final answer? His own personal preference will set the Army policy for Ft. Bragg? The other reason given by Fr. Whalen in his response, although considerate for Bishop Gossman, does not apply because my husband spoke personally with Bishop Gossman and he does not care if we have a Latin Mass here at Fort Bragg as long as it doesn't require a priest from his diocese.

I find this all so sad—we have met so many nice single soldiers and young families who also desire the Latin Mass here at Fort Bragg. None of us can understand the reason for these denials. In addition, a couple of months ago the *Army Times* ran an article on the Chaplain's Corp, and a chaplain boasted that he had intervened with a commander on behalf of a woman who wanted a Wicca service and he helped get it approved. This same chaplain wrote us a letter saying that the question of the Latin Mass was up to the chaplain, period. He offered no help. A priest here at Ft. Bragg told my husband that if they offered the Latin Mass here, they would also have to offer Wicca worship. He was wrong—the Army will allow Wicca worship but no worship in Latin.

Archbishop O'Brien, we are not asking for the Latin Mass in place of the New Mass. I grew up with the New Mass, I was married in a Nuptial Mass, and all my children were baptized in the new rite. I think that the two Masses can go beautifully together. One stresses the community, the other stresses the sacrificial. One stresses simplicity, the other stresses the divine. At Ft. Bragg there are other reasons as well to offer the Latin Mass. We are a very diverse population here. All could worship together without feeling left out. [Here she points out that there is Mass in English and Spanish, but no such provision for the Korean, Philipino, and other linguistic populations.]

Please advise me. If the answer is no, and that Fr. Whalen's feelings and decision are final, I will accept it with much sadness. If you think some type of solution or compromise can be worked out, I will wait and pray patiently.

I'm sorry this letter is so long. I would love to discuss any or all of this with you at any time. Thank you for your time. May our Father in Heaven continue to bless you and your work.

Sincerely,

Lorri L. Sonnier

Although my wife and I had written separately to the archbishop, he replied to both of us in the same letter:

ARCHDIOCESE FOR THE MILITARY SERVICES, USA

Telephone: (301) 853-0400
Fax: (301) 853-2246

Post Office Box 4469
Washington, D.C. 20017-0469

Office of the Archbishop

September 2, 1997

Major and Mrs. David L. Sonnier
105 Frazier Court
Fort Bragg, NC 28307

Dear Major and Mrs. Sonnier:

Thank you for your materials received two weeks ago and for the generous and understanding tone of your correspondence. Thank you for your call last week during my vacation.

I have reviewed the material you have sent as well as the policies of the Archdiocese for the Military Services on this matter.

I regret to have to inform you that I will not overrule your local priest's decision. Father Whalen has long been a most respected priest chaplain as was Father Marceaux before him and I must concur with the prudent judgment of the pastor on the scene in such matters.

Hopefully - and this is my genuine prayer -- the energies to generously expended in pursuit of this request can be graced in the direction of the full Fort Bragg Catholic community.

In the Lord,

+Edwin F. O'Brien
Archbishop for the Military Services

EFOB/plt

cc: CH (COL) Frank J. Whalen

*Serving Catholics in the Army, Navy, Air Force, Marine Corps, Coast Guard,
V.A. Medical Centers and Government Service Overseas*

September 2, 1997

Dear Major and Mrs. Sonnier:

Thank you for your materials received two weeks ago and for the generous and understanding tone of your correspondence. Thank you for your call last week during my vacation.

I have reviewed the material you have sent as well as the policies of the Archdiocese for the Military Services on this matter.

I regret to have to inform you that I will not overrule your local priest's decision. Father Whalen has long been a most respected priest chaplain as was Father Marceaux before him and I must concur with the prudent judgment of the pastor on the scene in such matters.

Hopefully—and this is my genuine prayer—the energies to [*sic*] generously expended in pursuit of this request can be graced in the direction of the full Fort Bragg Catholic Community.

In the Lord,

+Edwin F. O'Brien
Archbishop for Military Services

At this time all I could do was feel sorry for him. There was nothing more I could tell him. He had no way of knowing me, any more than he knew these two priests. I had given him all the information he needed to make the decision. He would have to answer to God.

Later that week, Lt. Gen. Schoomaker struggled with the issue for nearly an hour while I sat in his office. He had a seemingly infinite number of intelligent and relevant questions that anyone with a good heart would ask. He commented on how he had once seen a priest playing a banjo during Mass, and on how many Catholics he knew who no longer practiced their faith. In the end, he knew he couldn't help either. There was no way that a general officer who happened also to be a Protestant could dictate policy to Catholic chaplains. What we were requesting was only allowed by "indult," or by exception. The "Catholic" chaplains unanimously didn't want to do what the pope had asked them to do, and they had a convenient way to get out of it. The bishop of Raleigh, who had nothing at all to do with the Army, refused to allow the Tridentine Mass in his diocese, so they would follow on the pretext of "maintaining good relations with the Diocese of Raleigh." In our case, all the reams of Equal Opportunity policies didn't apply.

Any correction to the situation would have to come from the pope himself. For now, those Catholics at Fort Bragg who longed for the traditional Mass had few good options. However, by now we had discovered how to find the many private chapels scattered across the landscape. While driving home on vacation or visiting relatives or friends in other states, we would find a group of Catholics that had established a private chapel so that on one or more Sundays a month Mass would be held in a storefront, the conference room of a hotel, or even in someone's home. We attended a number of these in various locations while traveling. In each case, the diocese in which the private chapel was established was a diocese in which the bishop had not made provisions according to the guidelines of *Ecclesia Dei*. We timed our visits to Mississippi to be there on a weekend when Bishop Howze was offering the Latin Mass in the Cathedral in Biloxi.

Bishop Joseph L. Howze gave us the moral courage and strength to not be intimidated by the meanness and intolerance of the chaplains we encountered. He was quite unusual. He had grown up as a Protestant in the Deep South, but due to the excellent influence of Catholics around him he eventually embraced the Catholic faith and pursued his calling to the priesthood. After a series of assignments, he became auxiliary bishop in the Diocese of Natchez-Jackson in 1972, and then became the first bishop of the Diocese of Biloxi when it was created in 1977. Immediately after Pope John Paul II issued the 1984 indult *Quattuor abhinc annos*, allowing bishops to authorize the Latin Mass, he began offering the old Mass in the Cathedral. He then gave a universal indult for the priests in the Diocese of Biloxi. In a 1992 interview with *The Latin Mass* magazine he said, "I announced in my clergy bulletin that any priest who wanted could say it [the TLM] any other Sunday but the Sunday that I do, the first Sunday of the month. But...I haven't gotten one response."[1]

---

[1] *The Latin Mass*, March-April 1992, pp. 14-16.

He also told me this personally, adding that he was hoping that some Latin Mass communities would form within the diocese, and for that one Sunday every month they could all go to the Cathedral for a High Mass. He was disappointed that it never happened.

"I think the priests in the diocese didn't like me very much because I'm black," he once told me, after he retired in 2001.

"They didn't like you because you're Catholic," I muttered.

He also said that he was ashamed at the treatment of the "Latin Mass people" by other Catholics.

"It reminds me of the way our people were treated," he said.

The meaning was clear. It reminded him of the way blacks were treated in the Deep South when he was growing up. We often visited him during our trips home to Mississippi, even when we couldn't get there for his monthly Latin Mass. We continued to correspond with him long after he retired.

We also continued to periodically host a circuit priest who was willing to visit our residence once a month and offer Mass for one or two other families and a few servicemen. God finds ways to turn evil into something good. I began studying Gregorian chant on my own, with no guidance, and learned to play the "organ," which was just a keyboard that had organ settings. I had studied music only for a few years in my youth, but it came back quickly. Had a public Fort Bragg Latin Mass Community actually come together, I would have been too busy organizing altar boys and scheduling events to have discovered that I had been blessed with a gift of understanding music. We had the clandestine Latin Mass once a month, at most, but it always brought us a great sense of relief. When our fourth child, Louis, was born we had him baptized in the Maronite rite. As is the case in many of the Eastern rites, one receives the Sacrament of Confirmation at the same time, so he was confirmed as well as baptized. We invited the Bonomettis, our friends from Alabama, to be his godparents. The fact that this baptism included the rite of exorcism, as the preconciliar Latin rite did, was not lost on us. From that time on, the rest of our children would be baptized in the traditional Latin rite.

We sought refuge in the Maronite Rite. When our fourth child was born, we had him baptized at St. Michael the Archangel Maronite Catholic Church in Fayetteville, NC. Joe and Peggy Bonometti are the godparents.

With my father and Lieutenant General Pete Schoomaker at the ceremony in which I was promoted to Lieutenant Colonel.

A few months later I was promoted to Lieutenant Colonel. Lt. Gen. Schoomaker received his fourth star and went on to become the Commander in Chief, Special Operations Command (CINCSOC) at MacDill Air Force Base in Florida. Lt. Gen. Keane left about the same time, and eventually also pinned on a fourth star. The younger officers who had assisted me in circulating petitions and writing letters all left the Army over the next few years. One of them took the time to write a letter to Archbishop Edwin O'Brien at the Archdiocese for the Military Services:

> Dear Archbishop O'Brien,
>
> The purpose of this letter is to inform Your Excellency of my resignation from the Army due in part to the failure of the Catholic chaplains on Fort Bragg to support my preference for the traditional Mass. I have attended both the post and pre-Vatican II Masses celebrated in English and Latin. My preference by far resides with the Latin Mass. I have increasingly noticed a large number of young persons, both military and their dependents, who share such views.
>
> I have petitioned almost every Catholic chaplain on post, the former senior Catholic Chaplain COL Sydney Marceaux, Archbishop Dimino, and the newest senior Catholic chaplain on Fort Bragg, Father Whalen. In addition, I petitioned Bishop Joseph Gossman, of the diocese of Raleigh, NC. All of my requests, although supported by up to fifty signatures, have been denied or re-directed to another source.
>
> As you know Our Holy Father, Pope John Paul II granted the permission for the Holy Sacrifice of the Mass to be offered again in Latin. His Apostolic Letter, *Ecclesia Dei*, calls for the "wide and generous application of the traditional Latin Mass."
>
> Now I have resigned effective 31 Dec 1998 and I will move to another state or diocese which supports and acknowledges the Pope's instructions. In addition, I plan to vacation in Rome, and at such time I will request an office call with the necessary officials to make our dilemma known in person. I can be reached by writing or calling the above address and phone number. With these thoughts in mind, I remain

Respectfully yours In Domino,

Christian M. Neary
1LT(P), FA

Having declined consideration for battalion command, I accepted a position as an "Information Management Officer," essentially the Director of Information Technology, at North Atlantic Treaty Organization (NATO), Headquarters in Brussels. In the summer of 1998 Lorri and I left—now with four children—for Belgium.

Chapter 12

NATO HEADQUARTERS, BRUSSELS, BELGIUM

E WERE CAREFUL IN CHOOSING FROM THE options I was given. I was offered several positions in the U.S. and several overseas. When Brussels was mentioned I did a bit of research to be certain that I was moving to a region in which the local ordinary didn't appear hostile to the Tridentine Mass. Sure enough, the FSSP had an apostolate there.

Belgium is a country with a split personality. The northern half of the country, Flanders, is inhabited by the Flemish who speak a variety of Dutch that's not much different from the language spoken in the Netherlands. The typical Flemish adult also speaks several other languages. They're industrious, intelligent, and energetic. In the southern half of the country, Wallonia, the Walloons speak French. To outsiders this region seems to differ very little from France culturally and linguistically.

Belgium was once a Catholic country but, as in the rest of Europe, most Belgians lost their faith after the Second Vatican Council. At one time, virtually all Belgian Catholics met their Sunday obligation to attend Mass. When we arrived in 1998, Sunday Mass attendance had dropped to two percent of professed Catholics and was still declining.

We arrived in mid-June and found a house in the town of Overijse, located in Flanders but situated close to the French region of Wallonia. From downtown Brussels, and from the NATO Headquarters in Evere, it was a fifteen-minute drive, provided that drive was made in the middle of the night. During rush hour, it took typically 30–45 minutes. I found a bicycle route that could get me there in the same amount of time if the weather was good.

Within the U.S. Delegation to the NATO Military Committee, I oversaw computers, networks, network administration, telephones, future video-teleconference capability (which they had not purchased yet at the time of my arrival), and anything else that involved technology.

From the start it was clear that this would be a good assignment for me to either reinvent the rest of my career as a "computer guy" or finish off my army career and be done with it. Retirement was only three years away. I was still hopeful that the disorder in the Church would be corrected, Catholics in the military would be able to have access to the traditional Latin Mass in the base chapels, and there would once again be a purpose for my service. Although I had declined battalion command, which meant that I would almost certainly never be promoted to general officer, there was a possibility of continuing beyond twenty years.

The U.S. Representation to the NATO Military Committee is a three-star general position. His staff was relatively small. We were about 55 officers and enlisted personnel from various services as well as some civilian employees — mostly administrative. The environment was fascinating. All the other military delegations were located within the same building. Poland, the Czech Republic, and Hungary would soon join NATO in 1999, and some of those offices would be in a different building. The Military Representatives (MILREPs) from the various NATO countries met weekly (or sometimes more often) to discuss issues pertaining to NATO military policy. Each NATO country also provides an ambassador to NATO. The U.S. ambassador has by far the largest staff, and in 1998 it was at least three times the size of ours, and they had their own information technology staff.

Nearby, there was a small U.S. military complex, the NATO Support Activity (NSA). This complex, commanded by a lieutenant colonel, supported the 2,250 U.S. personnel assigned to Brussels. There was a finance office, a transportation office (to arrange for shipment of furniture, etc. to and from Belgium), and a variety of other administrative services. There was a shoppette — a sort of general store for servicemember needs. There was also a chapel and a U.S. Army chaplain assigned to the small complex. The chaplain at that time held the rank of major, and the position was normally filled by a Protestant.

For a family that was as dedicated to the Army as we had once been, we found it all quite amusing. People seemed to take themselves way too seriously considering that NATO had lost its *raison d'être.* NATO was originally established as a trans-Atlantic alliance to assure smaller European countries that they would not be abandoned should the Soviets roll across Western Europe. The NATO constitution included Article 5, which stated that an attack on any NATO country would be considered an attack on all, obligating all NATO members, the U.S. included, to pitch in for defense.

With the breakup of the Soviet Union in 1990 and 1991, the threat of a Soviet attack no longer existed. The various countries that had been part of the Soviet Union began making cautious advances toward better relationships with the West. Russia, depleted by nearly a century of trying to create a workers' paradise, was struggling to survive. Even though the Secretary of State, James Baker, promised that NATO would not expand eastward, we had broken that promise in less than a decade and were adding Poland, the Czech Republic, and Hungary. This was the first round of NATO expansions that would eventually lead to a dangerous confrontation with Russia.

Why had the U.S. continued to remain part of this alliance? In 1991 it would have been possible to dissolve NATO altogether. Alternatively, the U.S. could have left the alliance and let the other member nations manage themselves, now that there was no Soviet Union. Instead, NATO redefined its mission to commit to regional security and stability, but with a heavy reliance on the U.S. Since the mid-1990s some difficulties had been festering in the former Yugoslav republics and NATO was suddenly all over it.

Meanwhile, I set my sights on doing my information technology job and taking care of my family. One of my first duties was to travel to London for a few days of temporary duty to take a class. I contacted Michael Davies to see if he was available during the time that I would be there, and we arranged to meet at a train station a block or two from his home in Bromley, Kent. There I spent an afternoon with him in a surprisingly small room upstairs, where he had done all of his writing. We drank beer and watched a rugby rerun. He had the enthusiasm of an Alabama football fan, even knowing how the game would end. I got him to autograph a book he had recently published on the life of St. John Fisher.[1] As we walked back to the train station and he saw me off, a thought occurred to me. This ordinary-looking man

---

[1] Michael Davies, *Saint John Fisher* (Neumann Press, 1999).

standing at the platform was accomplishing more meaningful work on behalf of mankind than all the well-paid generals at the NATO round table and their secretaries and staff members combined. The pride and vanity of the increasingly bombastic coalition would one day fade away and this simple-looking man, currently known to very few, would be widely recognized as a pioneer in the effort to keep the traditional Catholic faith, doctrine and liturgy alive in the English-speaking world.[2]

---

[2] A full-length biography has finally been published: Leo Darroch, *Michael Davies, The Great Defender of Catholic Tradition* (Arouca Press, 2026).

CHAPTER 13

QUO VADIS?

URING THE FIRST YEAR WE WERE THERE, WE got to know both of the FSSP priests assigned to Belgium. They held a catechism class on Tuesday evenings at their residence, and I made it a point to be there as often as possible. I had the double pleasure of learning about our Catholic faith and honing my weak French skills.

The year was 1998, and the tenth anniversary of *Ecclesia Dei* was just around the corner. A series of activities were scheduled for the weekend of October 23–26 in Rome to commemorate the occasion. I thought that this would be a good opportunity to speak personally with the someone in the *Commissio Ecclesia Dei* about the predicament we were in. It seemed that someone in that office should be able to make phone calls, provide information, talk to bishops, and help the bishops to understand that words spoken and published by the pope required obedience.

I contacted Michael Davies, and he helped me to arrange an appointment. I would have a meeting with Msgr. Arthur Calkins on the first day of the anniversary festivities. I had never heard of him because my correspondence had been answered by Msgr. Camille Perl. Michael Davies told me that Msgr. Calkins was an American.

"What should I say?" I asked him.

"Say whatever you want to; just explain the situation."

"Should I wear my uniform?"

He laughed. "No."

I brought the entire stack of letters, correspondence, a few notes I made to myself, and rang the bell outside Msgr. Calkins's office. He met me at the door with Michael Davies. Msgr. Calkins spoke Italian to the secretary as we passed her on the way to his office. I had to wonder how long this American had been here, and why it was that Msgr. Perl had been the one to answer my correspondence rather than him.

As I started talking, he interrupted.

"I'll listen to what you have to say, but there's nothing I can do about the situation in Montgomery."

"Montgomery? I came to talk to you about Fort Bragg," I said.

"Alabama?"

"No, North Carolina."

"Oh. Go ahead. There's still nothing I can do but go ahead anyway."

I explained the situation in as few words as possible: that on the military installations we were always being told that the policy of the surrounding diocese had to be respected, but if you then write to the diocese, they refer you back to the military chaplaincy. I pointed out that this had happened both at Maxwell AFB in Montgomery, which is in the Archdiocese of Mobile, and at Fort Bragg, which is in the Diocese of Raleigh.

Msgr. Calkins had once been a priest in the Archdiocese of New Orleans. He had been there in the 1970s, and his recollection of parish life was the situation that existed in the 1970s. It was clear from the start that we had no common framework for understanding the problems that existed for a Catholic parent in 1998 trying to raise children and pass on to them the same Catholic faith we had received. When I spoke of the scandals of the Novus Ordo, he thought I was exaggerating. But then suddenly, he admitted that he had heard it all before and he believed me. But he thought I should accept it as it is instead of trying to restore the old liturgy. The source of his indifference or hostility to those of us who were making these requests for the Tridentine Mass to our bishops was never clear to me, but it was clear that he didn't like dealing with the matter. He seemed more interested in getting me to accept things the way they were and stop making issue with the liturgical problems. He even tried to convince me that the FSSP should begin offering Mass in the new rite.

"Perhaps if the FSSP were to meet people's expectation of a priest, in the way that they offer the Mass, they would be accepted in the military."

I had to wonder why in the world he had been assigned to this job. If he had been more energetic, answering letters with honest concern, paying visits to the SSPX seminaries, trying to coax them into discussions with the bishops, trying to persuade bishops to make concessions, he could probably make a difference. If he had shown a bit more interest in the situation, things could have improved years ago.

"There's not enough support to have a Latin Mass in a small rural Southern town," he said. I pointed out that there were over 40,000 servicemen and families at Fort Bragg and Pope AFB, and that many of them were not Southerners. Many of them were lapsed Catholics, and some of them could easily be retrieved by the presence of an orthodox Catholic community centered on the Latin Mass. I pointed out that there were 400 formerly Catholic families attending one of the local Protestant churches in nearby Fayetteville.

"We don't interfere in the decisions of the local bishops," he simply stated. I had to wonder what his job was, in that case! The bishops were at least partially responsible for the persisting problems. In both cases they had groups of people within their dioceses who had established chapels outside of their authority to worship in the traditional manner. The very existence of such groups affirmed the need to respect the guidance of the Holy Father, and yet Msgr. Calkins wasn't going to call anyone, clarify, explain — even write a simple letter?

Sensing that I was not impressed with his indifference to the struggle, he tried to appeal to my non-existent pessimism. "This will all be lost," he said, waving his arm in the general direction of St. Peter's Square. "All of it."

"It doesn't have to be," I responded, wondering why he said this. Perhaps he had noticed in his past conversations with tradition-minded Catholics that they tended toward pessimism, and maybe he had been able to find some common ground with them by appealing to it. This attempt to appeal to some assumed pessimism in my personality left me disturbed and disgusted.

He rambled on for a few minutes, during which he made some reference to *Christian Order*[1] as "Christian *Disorder*," and admonished me to attend the new Mass and offer up my sufferings in doing so. At least he acknowledged that there was suffering involved.

---

[1] *Christian Order* (www.christianorder.com), by its own description "a British international monthly devoted to the defense and propagation of the Catholic, Apostolic and Roman Faith — through incisive comment on current affairs in Church and State, both at home and abroad."

After our unnerving meeting I had a blistering headache. This had been one of the most disappointing episodes so far. Michael Davies urged me to be patient.

"Actually, he's come a long way," he encouraged me.

"Oh, really? How so?"

"It's because of him that the Society of Saint John now exists in the Diocese of Scranton."

The Society of Saint John was a group of priests who had been with the SSPX only a year or two before. Under the helpful guidance of Bishop Timlin of the Diocese of Scranton, Pennsylvania, they established a new society that had about 15 members. This organization would last only a few more years, before being removed from the Diocese of Scranton by Bishop Timlin's successor, Bishop Joseph Martino, under rumors of financial and moral corruption. However, Bishop Timlin had also allowed the establishment of the FSSP Seminary in his diocese, which would eventually relocate to the Diocese of Lincoln, Nebraska. He also supported an order of nuns founded in 1995 by Sr. Wilhelmina Lancaster, the Benedictines of Mary, Queen of Apostles, who would eventually relocate from Scranton to Gower in the Diocese of Kansas City, Missouri, around 2006. At that time, the fact that Bishop Timlin had made a home for multiple groups attached to traditional practices of the Catholic faith set him apart from most of the other bishops throughout the world.

Bishop Timlin had been a friend to people who loved the old Mass from the early 1990s, and he had taken such interest in the efforts of the FSSP priests that he also decided to travel to Rome for the tenth anniversary of *Ecclesia Dei*. He was currently in Rome, and later that day I had the opportunity to meet him. He was offering Mass at the North American College. I arrived early and sat near the front as he entered the chapel and Mass began. I was suddenly quite emotional. Why couldn't more bishops be like Bishop Timlin? He had done so much, and yet it probably hadn't cost him that much in time and effort. All he had done was to keep his mind open to what the Holy Father had said, and leave doors open to those who desired to follow the old rite in his diocese.

I met Bishop Timlin after Mass and introduced him to some priests from Spain who fell into the category of "independent traditionalists," meaning that their bishops in Spain did not recognize their order. I translated as Bishop Timlin spoke with their superior, until they discovered that they both spoke Italian.

Bishop Timlin celebrated a Pontifical Mass on Saturday in the beautiful church of Santa Maria della Scala in Trastevere. I think he was shocked by the large throng that met him outside the Church. Catholics from all over the world crowded around him, thanking him for all he had done, and treated him quite literally like a rock star. Yet all he had done was to be a good shepherd.

During the busy three-day period, we were addressed by the Holy Father, who had this to say at a papal audience on October 26:

> I invite the Bishops also, fraternally, to understand and to have a renewed pastoral attention for the faithful attached to the Old Rite...

Cardinal Ratzinger, the future Pope Benedict XVI, gave a lecture at the Ergife Palace Hotel that Saturday in Rome, to an audience of about 3,000 Catholics. He had this to say:

> The Council did not itself reform the liturgical books, but it ordered their revision, and to this end, it established certain fundamental rules. Before anything else, the Council gave a definition of what liturgy is, and this definition gives a valuable yardstick for every liturgical celebration. Were one to shun these essential rules and put to one side the *normae generales* which one finds in numbers 34–36 of the Constitution *De Sacra Liturgia* [i.e., *Sacrosanctum Concilium*], in that case one would indeed be guilty of disobedience to the Council! It is in the light of these criteria that liturgical celebrations must be evaluated....

Or, in other words, those who had eliminated Latin from the liturgy were guilty of "disobedience to the Council"! But why was it that these words of the pope and Cardinal Ratzinger never seemed to translate into practice at the diocesan or parish level in a way that would make life easier for loyal Catholics? Hearing this from Cardinal Ratzinger reaffirmed that it mattered not what the USCCB thought of me; *they* were the ones guilty of disobedience. The cardinal and future pope continued:

> We must now examine the other argument, which claims that the existence of the two rites can damage unity. Here a distinction must be made between the

theological aspect and the practical aspect of the question. As regards what is theoretical and basic, it must be stated that several forms of the Latin rite have always existed, and were only slowly withdrawn, as a result of the coming together of the different parts of Europe. Before the Council there existed, side by side with the Roman rite, the Ambrosian rite, the Mozarabic rite of Toledo, the rite of Braga, the Carthusian rite, the Carmelite rite, and best known of all, the Dominican rite, and perhaps still other rites of which I am not aware. No one was ever scandalized that the Dominicans, often present in our parishes, did not celebrate like diocesan priests but had their own rite. We did not have any doubt that their rite was as Catholic as the Roman rite, and we were proud of the richness inherent in these various traditions. Moreover, one must say this: that the freedom which the new order of Mass gives to creativity is often taken to excessive lengths. The difference between the liturgy according to the new books, how it is actually practiced and celebrated in different places, is often greater than the difference between an old Mass and a new Mass, when both these are celebrated according to the prescribed liturgical books.[2]

So much for the accusations of "dividing the Church" and "causing disunity." Most Catholics wouldn't know the difference between the Novus Ordo Mass, properly celebrated with a reasonable retention of Latin, and the Tridentine Mass. All we could do was hope and pray that this cardinal would one day be pope.

We were also addressed by Dom Gerard, the Abbot of Le Barroux (a Benedictine monastery in southern France) and Michael Davies. When Davies saw me in the audience, he came over, leaned to my ear, and with characteristic humor said: "Whatever you do, don't abandon your chair for an instant or you will return to find a Frenchman sitting in it!" The French had indeed turned out in large numbers for the occasion.

I came away from the weekend understanding the situation much better than when I went into it. I was more resolved to fight, but at the same time I could see that the root of the

---

[2] Joseph Cardinal Ratzinger, "Ten Years of the Motu Proprio *Ecclesia Dei*," *Adoremus*, December 31, 2007, https://adoremus.org/2007/12/ten-years-of-the-motu-proprio-quotecclesia-deiquot/.

problem was simply human weakness and cowardice among the U.S. bishops. Most of them were unable to resist ideological pressure, so they were running away and leaving the sheep to fend for themselves. There was nothing we could do about it but to pray—but it became much more urgent to me that our prayers should be the same as the prayers of those who came before us. Michael Davies told me a story about a Catholic family from France that had recently spent two weeks vacationing in England before they realized that the church they were attending was Methodist. They could not tell the difference. I made it my goal that within my family we would live our lives so that if a Catholic from a previous century were to time-travel into the present, he or she would be able to identify us as authentic Catholics. We would keep the same Mass, prayers, devotions, music, and culture as much as possible.

A few months later, during Pentecost weekend of 1999, I took another weekend trip, this time to make the pilgrimage from Notre-Dame de Paris to Notre-Dame de Chartres. This annual three-day pilgrimage is the largest gathering in the world of Catholics who follow the traditional rite. By this time in the pilgrimage's modern history, the third day of the pilgrimage brought together about 15,000 Catholics crowded in and around the Chartres Cathedral.[3] I brought my oldest son, and we walked with the Belgian chapter the first two days, and with one of the American chapters on the final day.

Most of the participants in the annual pilgrimage are French. At the end of the pilgrimage I was convinced of two things: that the French, generally speaking, are not at all like the Parisians one meets in the capital city, and that despite the best efforts of the revolutionaries, the Latin Mass will not die.

---

[3] In recent years, this number has swelled to around 19,000 (2025), and it would be even greater if the coordinators were not forced to put a cap on registration for logistical reasons (the areas and resources they have available simply cannot accommodate any more bodies!).

CHAPTER 14

THE BIG LIE

A T SOME POINT DURING THE SUMMER OF 1999 Lt. Col. Don Isbell took command of the NATO Support Activity. Don was a friend and colleague, and Don's wife Emmy had worked with Lorri in the Cape Fear Valley Regional Hospital in Fayetteville during our first tour at Fort Bragg. Now that they were together again, they often took early morning walks along the quiet residential streets of Overijse. Don was a good commander and was doing his best to succeed at his new post. The NSA chaplain under his supervision was a Protestant. In addition to counseling troubled servicemen or their families, he held a non-denominational Protestant religious service on Sundays in the NSA chapel. He also did a variety of other things as a staff officer.

It turned out that one of the noncommissioned officers who worked in the personnel office down the hall from me at NATO headquarters had been a preacher before coming into the army. At his request, a second religious service, a "Gospel Worship Service," was established at the NSA chapel, and this sergeant first class was placed in charge of it. So now, there were two Protestant religious groups meeting at the NSA Chapel: one led by the Protestant chaplain, and the other led by a soldier who had once been a Protestant pastor in a previous life.

"It was easy," Don said. "We pushed the paperwork through, and it happened in no time at all. Maybe a month or two."

This brief conversation piqued my interest. Under different circumstances I would probably have left things alone. Now, my colleague Don was able to make a new religious service at the NSA chapel happen, and he was describing the process as "easy." The situation that had previously caused us problems no longer existed. The NSA was located within the geographical limits of an archdiocese in which one could find a traditional Latin Mass, so having a Latin Mass in the NSA chapel couldn't be opposed on the basis of "respect for the policies of the local bishop." The problem we had encountered at Fort Bragg was that the bishop of Raleigh had been opposed to the use of Latin in the liturgy or the restoration of the 1962 *Missale Romanum*. And the military chaplains, out of "respect" for the Diocese of Raleigh, claimed the right to reflect the same policies within the military base and they were only too happy to exercise it.

In Brussels, the local ordinary was Cardinal Godfried Danneels. What would the military chaplains do if we were to initiate a request to use the NSA Chapel for a weekly Tridentine Mass? They would not be able to say, as they had done at Fort Bragg, "No, the local bishop doesn't allow it so we can't either." They would not be able to claim that there was not a priest available. We met several very good priests during the first year we were in Belgium who spoke English perfectly well and followed the old rite. The FSSP had two priests assigned to Brussels.

I asked Don if he would assist us in establishing a Latin Mass community. It wouldn't cost a cent, since I would be glad to cover any costs of bringing our priest there. I was fully prepared to buy whatever was needed, such as missals or hymnals.

"You don't have to do that! There are funds available for that," Don assured me. He was excited about the idea. "I used to be a Catholic back when the Mass was in Latin," he added.

Soon we presented the idea to the Deputy MILREP, who was Air Force Brigadier General Earnie Callender. In addition to being the Deputy MILREP, which meant that he had to attend the NATO Military Committee meetings when the MILREP was unavailable, General Callender oversaw all U.S. Community Support activities. In this capacity, he was Don's boss, and he took this responsibility seriously. While he operated "by the book," he also kept an eye toward making life as livable as possible for military families assigned to Brussels.

I was eventually drawn into a meeting with him about having a Latin Mass at the NSA Chapel. He asked a few questions, then told me about the process through which he had recently established the "Gospel Worship Service" there. He was rather surprised that there were priests around who still offered the Latin Mass. He was even more surprised when I brought Fr. Gérald Duroisin, FSSP, in to meet him. Fr. Duroisin, at the time, was around forty years old. I think General Callender was expecting to meet an elderly priest in his seventies or eighties.

General Callender confirmed that Fr. Duroisin's command of the English language was sufficient for him to be involved in the U.S. military facility in this capacity. The two of them had a pleasant discussion. Fr. Duroisin had a knack for connecting with people. A couple of the secretaries kept him for a few extra minutes as we made our way out of the office. He was a very dedicated and serious priest, and anyone who spent more than a few minutes with him could immediately detect that he was a man of God.

Our Staff Judge Advocate, Lt. Col. Albert Klein, drafted a memorandum. He seemed to know everything about everything. He had a mind like an encyclopedia, and within two weeks of our initial conversation on the subject, he had already researched canon law, provisions made for the Indult Mass in various locations throughout both the U.S. and Belgium, the history of the Latin Mass movement, and a variety of other matters related to our request. And he had done all of this in the middle of handling his regular job, which involved difficult legal issues pertaining to NATO, environmental law in Europe, and a variety of other legal problems. He was extraordinary.

The memorandum, dated August 5, 1999, was a major accomplishment in that it came from a general officer and therefore would have to be responded to, in writing, by a general officer:

> Office Of The United States Representative Military Committee
> North Atlantic Treaty Organization
>
> Memorandum Thru: 80th Asg, Attn: Chaplain (Col.) Sydney Marceaux
>
> Memorandum Thru: Hq USAREUR, Attn Chaplain (Col.) James J. Jagielski
>
> For: The Archbishop For The Military Services
>
> Subject: Religious Service at NATO Support Activity Chapel
>
> 1. I am writing to enlist your assistance in establishing a traditional Catholic Mass in the Brussels American military community. The Brussels American community includes over 2,000 military, civilians, and family members from numerous U.S. government and NATO activities to include NATO Headquarters, the U.S. Embassy, and the U.S. Mission to the European Union. The NATO Support Activity (NSA), 80th Area Support Group, is responsible for providing chaplain services to this community.

2. Three months ago, the NSA moved into a newly constructed building, which includes the first-ever, dedicated chapel complex in our community's 32-year history. The complex includes a co-located chaplain's office, chapel, and fellowship room. It represents a quantum leap in the ability of our leadership to provide the community with appropriate religious support.

3. Historically, this community has been served by Protestant religious services in the NSA Chapel, even before completion of the new facility, but to the best of my knowledge we have never held Catholic services there.

4. As the Community Coordinator, I recently received a request from Catholics in this community for a regularly scheduled traditional Mass. Catholics in our community often accommodate themselves with one of the various English-speaking congregations in the area, which follow the modern rites of the Catholic Church. The traditional Mass is available to French and Flemish-speaking Catholics in Brussels, with the permission of the local Bishop, but there is no traditional Mass available for English-speaking residents.

5. The request from community members was followed by an office call with Fr. Gerald Duroisin, of the Fraternity of Saint Peter (FSSP). I was impressed with Fr. Duroisin's sincerity in his desire to offer the traditional Mass to our community, as well as his excellent command of the English language. I have always associated a high Quality of Life with choice, and I believe this community would be well served by having the opportunity to participate in a religious service that is not currently available.

6. I understand we must coordinate this request through your office before holding services in our new chapel. I am forwarding this request through the 80th ASG and HQ USAREUR for your final consideration and approval.

On behalf of the Brussels American community, thank you in advance for your consideration in this matter. Please feel free to call me if you would like to discuss this initiative further.

M. E. CALLENDER, JR.
Brigadier General, USAF
Community Coordinator

The observant reader will notice a familiar name: Fr. Sydney Marceaux. After leaving Fort Bragg he was assigned to Supreme Headquarters, Allied Powers in Europe (SHAPE) at Mons, which was about a 45-minute drive southwest of Brussels. Fr. Marceaux, in his position, was the only Catholic priest assigned as a U.S. Army Chaplain in Belgium. There were not many U.S. military assigned there, and he technically had responsibility for all Catholics assigned to Brussels. He had never visited the U.S. Catholics in Brussels as far as anyone could remember. He had never concerned himself about where Catholics in Brussels attended Mass, or what went on in Brussels. I doubt he knew that I was assigned there until he saw this letter.

Most U.S. Catholics in Brussels attended Mass at St. Anthony's, a nearby parish that had an Irish priest and an international community of English, Irish, Americans, and a few English-speaking Africans. The liturgy there had taken the same horrible turns over the last years as had most modern parishes, so I couldn't even think of taking my family there. The Sonnier family had become very attached to a rural parish in Cortil-Noirmont where there was an Indult Mass.

Suddenly, all of this was of the utmost importance to Col. Marceaux. He immediately dropped whatever important work he was doing and made a quick trip to Brussels. Within two hours he was standing in the general's office.

Since the beginning of my assignment to NATO, I had been trying to live somewhat like the Europeans. I knew I would never do it, but at least I could try. I tried having a glass of wine with lunch once but found that I couldn't focus on my work that afternoon, and half my day was a complete loss. But regardless of what was happening, I did take about ten minutes every afternoon to boost my focus with an espresso in the NATO cafeteria. The place was always packed. I would visit with some other American who had the same idea, or some European colleague I recognized, or just sit for a few minutes and read *Stars and Stripes* or the *Early Bird*, which was a compendium of noteworthy articles from U.S. news sources.

On this day, as Fr. Marceaux rushed in for an emergency meeting to counter the Tridentine threat, I spotted him checking in with security guards. He obviously hadn't visited this building before, since he came through the wrong door. They gave him directions in French, and he responded in French. In fact, his French was quite good. I was tempted to stop and ask if I could help him, but I was also concerned that this would arouse his anger or have some other unpredictable consequence since he didn't seem to be rational when it came to issues involving the Latin Mass. I avoided attracting his attention, and just watched from a few feet away, comfortable that he would not recognize me wearing a business suit.

I enjoyed my espresso and returned to my office. About an hour later, Gen. Callender called me to his office.

"This may be harder than I thought," he said.

I wasn't surprised, but I pretended to be.

"Why?" I asked.

"I was just visited by none other than Col. Marceaux. He's very much opposed to this, and he remembers you from Fort Bragg. He tried to tell me about Fort Bragg, and I told him that I didn't want to hear it; that this is a different time, a different place, and a different assignment."

"Well, do you want to know about that, sir?"

"I'll tell you the same thing I told him: No."

Silence.

"I also caught him in a lie," he finally said.

Now this was interesting.

"He said that you and Fr. Duroisin are with that excommunicated French archbishop Le . . . whatever his name is. You know who I'm talking about. At first, I believed him, and I didn't care, I just thought, 'well, so what,' but then he spilled the beans. He said that Fr. Duroisin's order is the Society of Saint Pius."

I was intrigued. "So how did you know Fr. Marceaux was telling a lie?"

He leaned over on his desk and lowered his voice. "Fr. Duroisin told me that his order was the Society of Saint *Peter*, not Saint *Pius*."

He jumped up from behind his desk and began pacing the room; he had a tendency toward being animated when he was engaged.

"So I said to him, 'No, Col. Marceaux, it's not Saint *Pius*, it's Saint *Peter*. I have it written down right here in the notes from my meeting with him.'"

I was amazed. Now here was a general! There are many men with physical courage, but to have the moral courage to challenge a chaplain who is obviously lying seemed to be such a rare thing these days.

"So, then what, sir?"

"So then, Col. Sonnier, he began backpedaling. He began stuttering. He said something like, 'Well…well…, what I meant to say was, well…they have the *spirit* of the Society of Saint Pius.' But I knew better. I knew I had just been lied to by a priest. A Catholic priest!"

This was not good. We all look bad when our priests tell lies. But at least Fr. Marceaux had been caught, and General Callender knew what was going on. I had given Fr. Marceaux plenty of information about the Priestly Fraternity of Saint Peter when we were at Fort Bragg. I had given him newsletters and brochures about their seminary and their various apostolates. I had made it very clear to him that I was trying to play by the rules and establish a Latin Mass community within the jurisdiction of the visible Church structure.

"So, what do we do now, sir?"

"We wait. He said that he must get permission from the Military Archdiocese. We'll wait and see what they have to say."

CHAPTER 15

LETTERS ACROSS THE ATLANTIC

A FEW WEEKS LATER I RECEIVED AN EMAIL from Don; it was a courtesy copy of an email sent by Fr. Marceaux to Gen. Callender:

Sir,

Before leaving, as you are aware, I requested a meeting with Cardinal Danneels. When I returned, there was a letter from his Vicar General (i.e., Deputy Commander) stating that the Cardinal does not support the request. Now that I know the Cardinal's feelings, I will include it in my memorandum for the Archbishop for Military Services. I plan to have this, along with your memorandum, in the mail by COB Friday 3 August. Please let me know if you have any questions. Thanks for your patience in this delicate matter.

Sidney J. Marceaux
Chaplain (COL)
80th ASG Staff Chaplain

A delicate matter indeed! I had to wonder exactly how and in what manner he had presented the request to Cardinal Danneels.

A few weeks later Gen. Callender called me into his office. Walking around his desk, he took a seat across from a small coffee table and proceeded to tell me the bad news.

"I received this email from our friend in Mons, and he says the cardinal doesn't want us to have a Latin Mass."

"Oh?"

"But, based on my own conversation with Col. Marceaux a few weeks ago, I'm not sure I trust him."

"I'm not sure I trust him either, sir."

"He gave me this…" he said, handing me a photocopy of something that looked like a page out of a regulation. It appeared to be guidance to military chaplains:

> 1. The Tridentine Mass — A letter from the Congregation
> for Divine Worship dated 3 October, 1984 authorizes the
> bishop to permit the Tridentine Mass under certain
> conditions when he is petitioned to do so. Such a Mass
> should not take the place of any normal Liturgy that is
> regularly scheduled but it should be offered at intervals
> determined by the bishop for a specific group of peo-
> ple. Such groups should not be composed of those who
> impugn the validity or correctness of the revised liturgy.
> The Mass must be in Latin, with no interchange of texts
> or rites from the new Missal. Full details about the cir-
> cumstances should be sent to the Archbishop, who will
> then determine to what extent such a request can be
> accommodated. Such a Mass at any installation should
> not be in opposition to the policies of the local diocese,
> lest the Sacrament of Unity be a cause of division.

"Sir, this is outdated. This letter is based on *Quattuor Abhinc Annos*,[1] which was issued in 1984. It says nothing about *Ecclesia Dei*, issued in 1988. Also there has been more recent guidance."

Lt. Col. Klein, the JAG officer, walked into the office just then.

Gen. Callender continued: "I'm not sure that he didn't run to the cardinal and get the answer he wanted to tell us. Anyway, he said that permission had to come from the Military Archdiocese, so that's who I wrote to. And that's who will answer me."

I couldn't yet see where he was going with this. I was accustomed to getting denials and seeing people in positions of responsibility just give up.

He stood quickly and began pacing. "I am *not* in the habit of writing to someone, whoever that someone may be, and getting a response in the form of a poorly typed email from someone else. I wrote to Archbishop O'Brien, in Washington D.C. I got this response from Col. Marceaux in Mons, Belgium in the form of an e-mail! Is that how you Catholics do things?"

I had to laugh. I wanted to tell him that, apparently, it was how we did things when requests for the Tridentine Mass were on the table. Mirroring my thoughts, he continued in a low voice:

"I'm not sure my letter ever got to Archbishop O'Brien. These 'consultations' took place with Msgr. Callaghan, whoever that is, and he says right here..."

---

[1] Congregation for Divine Worship, *Quattuor Abhinc Annos*, October 3, 1984. The text may be found online at https://lms.org.uk/quattuor-abhinc-annos.

As he looked at his computer I had to wonder whether Msgr. Callaghan had ever given Archbishop Dimino the earlier requests we had sent from Fort Bragg. It had been the same situation there; my letters to Archbishop Dimino were always inadequately responded to by letters from Msgr. Callaghan.

"...Here it is. It says that the request is not appropriate since the local ordinary opposes it. This came to me this morning as an e-mail! So, I send a request to Archbishop O'Brien and what do I get in return? An e-mail from someone named Callaghan I've never heard of or met, forwarded to me by Marceaux in Mons!"

He slapped the papers on the table. "Nowhere do I see anything about what Archbishop O'Brien thinks of our request."

I was sure by now that they had hidden the whole thing from the archbishop. Good thing a general officer was involved in this. Or was it? In fact, it was embarrassing to see the dishonesty of these men in the Catholic hierarchy laid bare.

"I'm drafting another letter with the assistance of my lawyer," he nodded toward Lt. Col. Klein. "I'm not sure what it will say, and I'll send it along with our previous request, to his private fax number which my lawyer will find for me."

It would be a few weeks before Gen. Callender would send that letter to Archbishop O'Brien. In the meantime, I decided that there was nothing to stop me from sending a personal letter myself, expressing my concerns about the way it had been handled:

22 October 1999

From: Office Of The United States Representative
Military Committee
North Atlantic Treaty Organization

To: Most Rev. Edwin F. O'Brien
Archdiocese for the Military Services

Your Excellency:

Enclosed you will find a copy of correspondence Brig. Gen. Earnie Callender sent to you through the offices of the 80th Area Support Group (ASG) Chaplain and the USAREUR Chaplain requesting your approval for a regularly scheduled traditional Mass within our community. This Mass would be offered at the NATO Support Activity (NSA) chapel.

Two days ago, Brig. Gen. Callender expressed his concern to me. He asked whether or not it is typical in the Army, or in the Catholic Church, to send a letter to someone and to receive no letter in response, but rather an email from a different person. I assured him that you always answer your mail, and that perhaps his letter had not arrived in your office until recently, or that you were traveling and had not had a chance to respond.

Being satisfied with that, he began to read to me the contents of the email he had received from the 80th ASG Chaplain, Fr. Marceaux. What I heard was quite disturbing, and I must now express my concerns. The correspondence originated with Msgr. Aloysius Callaghan and had no reference to Your Excellency. It merely stated that the request was not considered appropriate since the local ordinary opposes it.

This is the same rationale used previously to deny petitions for the Tridentine Mass at Fort Bragg. If it were the case that the Belgian bishops were opposed to allowing the traditional Mass as the Bishop of Raleigh is, there would be some merit to this statement. To the contrary, however, Cardinal Danneels has allowed the Tridentine Mass to be offered in a number of places. It is rather common to find Latin Masses in both the old and new rite within Flemish-speaking Flanders, French-speaking Wallonia, and multi-lingual Brussels. The Fraternity of Saint Peter offers daily Mass during the week at a location close to the EU buildings, and on weekends they travel to other locations while diocesan priests offer the Tridentine Mass at several locations in and around Brussels. Currently there is no Tridentine Mass offered for the Anglophone community, and apparently Cardinal Danneels doesn't see us as his responsibility.

Your Excellency, Brig. Gen. Callender and I announced this initiative at the last Town Hall meeting, publicized it and had every expectation that it would receive your approval with ease. We simply cannot make a public announcement that the request is denied due to opposition by the local ordinary, because there is no such opposition. The Catholics interested in this initiative know that. They are familiar with and have attended Tridentine Masses offered in the French and Flemish speaking parishes; some attended the Notre Dame de Foy Pilgrimage. They just don't understand the sermons in French or Flemish and would like to have the same Mass offered in the NSA Chapel. To publicly announce the response Brig. Gen. Callender received via email would be confusing to the faithful.

I concluded the letter by inviting him to offer the first Mass for us, and informing him that we had an organist, a choir, and well-trained altar boys.

About a month later Gen. Callender received the following letter from Archbishop O'Brien:

November 18, 1999

From: Archdiocese For The Military Services

To: Brig. Gen. M.E. Callender, Jr.
Community Coordinator
Brussels, Belgium

Dear General Callender:

Thank you for your letter of 5 November which reached me two days ago.

I would like to grant the permission you seek for the traditional Latin Mass. I would not do so, however, unless the local bishop approves. I have a letter from Cardinal Danneels which states his desire that there not be a traditional Mass in English within his Archdiocese. I have written him for clarification in the hope he would reconsider.

Finally, while I know nothing of Father Duroisin, I presume he enjoys the faculties of Mechelen-Brussels. I have asked the Cardinal about this as well.

As soon as I hear from His Eminence, I shall be in touch with you. Meanwhile, thank you for your patience, your strong Faith and your interest. Do know of my prayers.

In the Lord,

✝ Edwin F. O'Brien
Archbishop for the Military Services

A few weeks later the response came through:

December 6, 1999

From: Archdiocese For The Military Services

TO: Brig. Gen. M.E. Callender, Jr.
Community Coordinator
Brussels, Belgium

Dear General Callender:

First, my thanks to you for patience in awaiting this response to your request for a Tridentine Latin Mass on behalf of some American military Catholics serving in the Brussels area.

The clarification I awaited arrived today in a letter from Godfried Cardinal Danneels. It leaves little doubt as to what my response to your request should be:

"The unanimous advice of my collaborators, is not to grant another permission, in order to avoid all kind of divisions in the Church in Belgium, based on liturgical preferences. This is also my personal opinion. Therefore, I would regret that such a permission be granted to the NSA chapel."

His Eminence offers further reasons for his position but in substance, he is quite definitively opposed.

I must, therefore, refuse your request. I hope that those who looked forward to this privilege will see the will of God working in spite of their preferences, and find the enrichment promised by the Lord in and through the sacraments of our Church.

In the Lord,

✠Edwin F. O'Brien
Archbishop for the Military Services

Gen. Callender called me in to his office to show me the letter. It was late in the day, and there were a million other things going on, but I think this situation disturbed him, as it would disturb anyone watching these events unfold. It didn't add up, of course. The pope had asked for generosity from "bishops and all those engaged in the pastoral ministry," the general had seen and read the key text of the document *Ecclesia Dei*, and yet there seemed to be such stubborn resistance. It wasn't a healthy, godly, holy resistance to something wicked, but precisely the other way around. It was a resistance based on bureaucracy, maneuvers, slander, and lies. He must have found it all quite revolting.

"Dave, this looks like bad news, but it's really not anywhere near over. It just means that I have to take a bit of my time at some point during the next two weeks. "

I couldn't say much at this point that would help. The rationale looked ridiculous; I had heard Cardinal Ratzinger and the Holy Father himself refute the notion that it would be somehow "divisive" to have a Latin Mass. What were these "divisions" Cardinal Danneels was referring to?

"Sir, did you know that Mass attendance is down to 2% on Sundays in Belgium?"

He pretended not to be interested in such statistics, but they're hard to ignore.

"Sir, it was around 80% a few decades ago, if not higher. I'm not sure I understand what the cardinal is referring to when he says that he wants to 'avoid all kinds of divisions in the Church in Belgium.' There's hardly any church left."

"That may be, Colonel, but it's irrelevant to my job here. I'm in the business of ensuring that everyone gets treated the same. I believe in 'equal opportunity,' and I think that applies to everyone. It appears that you, and your family, and your colleagues whoever they are — and I don't care to know who they are — are caught in the middle of some mischief."

"What can we do at this point?"

"You don't need to do anything. I'm going to pay a visit to the cardinal."

It would be a few weeks before that meeting took place. In the meantime, there was considerable panic in the technology sectors of the world as we approached January 1, 2000, the beginning of a new millennium. The panic wasn't about the mystical or spiritual significance, but the very real concern that much of the legacy software in use in 1999 had been written without taking into consideration that we would not always be in the 1900s. This panic, known as the "Y2K Crisis," caused some of us who worked in technology support to lose sleep. January 1, 2000, was fast approaching, and software written with only two digits representing the "year" field would represent the new year as "00," thus interpreting it as the year 1900 instead of the year 2000. This problem had its origins in the effort to save memory space in the early days of computing by representing the year with only two digits instead of four. We had to take steps to ensure our software was compliant and then wait to see what happened. On New Year's Eve, along with all of the other tech professionals in the building, I participated in a worldwide teleconference in which, at the top of each hour, every military installation in the world reported in as they reached the critical moment, either to report that all was well or to report which systems were failing. As it turned out, the remediation and preventive measures that had been taken were sufficient to ensure that there were no widespread system failures. Shortly after midnight we were sipping champagne.

On February 4, *Anno Domini* 2000, two blue-suited officers in the United States Air Force pulled up outside the residence of Godfried Cardinal Danneels in Mechelen, northeast of Brussels. The receptionist attempted to separate Lt. Col. Klein from the general. He politely but firmly remained with his boss, as the two were escorted into a large room.

"What is the purpose of your visit?" the cardinal asked coldly. As Gen. Callender would later recount the story, it was by far the coldest reception he had ever had from a "man of the cloth." He explained the request for regular Latin Masses to be held for Americans in the NSA Chapel. He clarified that the chapel was on a closed compound accessed only by ID-card holders and that nobody outside of the compound had to know anything about the Latin Masses held there. Finally, he detected a faint smile on the cardinal's face.

"So, that's it?" the cardinal asked.

"That's it," the general replied.

His Eminence asked the general to put the same request in writing and he would give it consideration.

It took five more months for his approval of the request to go through, but finally we received it in July 2000. Meanwhile, in April 2000 our fifth child, Annie, was born. Her godparents were Alexandra Colen, a member of the Belgian parliament whom we had met through the homeschool group, and none other than Lt. Col. Klein, who it turned out was a devout Byzantine Catholic.

Baptism of our fifth child, Annie.

CHAPTER 16

FINALLY!

FTER ONE YEAR OF APPEALS IN MONTGOMERY, Alabama, three years in North Carolina, and now two years after arriving in Belgium, *finally* we had succeeded! It was too good to be true! My confidence in the military had been restored. So what if this struggle had cost me so much in time and effort, having to both decline command and relocate. Now it was over. We had followed the shifting logic of denial to its final point, in the cardinal's office, and there it met its death. There could be no more denial of our requests from the hierarchy; those days were over. It had been well worth the struggle. In the end God's Church remained victorious.

The FSSP was under new leadership, and Fr. Duroisin was transferred to Lyon, France. We began looking for a priest who spoke English well enough to be comfortable offering our Latin Mass at the NSA chapel. The Institute of Christ the King had just assigned a newly-ordained priest from England to Havré, just outside of Mons, and it seemed likely he would be able to take responsibility.

Gen. Callender sent notifications to everyone he had spoken with previously: Archbishop O'Brien, the Chief of Army Chaplains, and the USAREUR and 80th ASG chaplains. As had become our custom, Lorri and I kept Gen. Callender in our Rosary intentions that evening. But almost as soon as we finished our prayers of thanksgiving, the saga took a new twist.

With Msgr. Gilles Wach and recently ordained
Fr. William Hudson at the residence in Havré.

Solemn High Mass at Notre Dame de Bon Vouloir in Havré.

CHAPTER 17

THE U.S. ARMY VS. THE CATHOLIC CHURCH

THERE WAS A MOMENT OF UNCERTAINTY. Although Archbishop O'Brien had seemed to indicate that all would be well as soon as we received permission from Cardinal Danneels, it was now unclear whether he had the ability to give such permission. Perhaps the question was whether he had the courage to give permission. My heart sank as we read the following letter sent to BG Callender from Cardinal Danneels:

> August 18, 2000
>
> Brig. Gen. M.E. Callender
> Office of the United States Representative Military Committee
> NATO-Brussels
>
> Dear General,
>
> The Archdiocese for the Military Services of the United States informed me that I didn't have the authority to grant the permission for the celebration of the Eucharist according to the rite of John XXIII.
>
> This privilege has to be granted by Archbishop O'Brien in Washington as the NATO-offices in Brussels belong to his authority.
>
> I regret that I wasn't aware of this canonical regulation. The permission I granted on July 7th wasn't valid. With my best wishes, I remain
>
> Yours sincerely in Christ,
>
> ✠ Godfried Cardinal Danneels,
> Archbishop of Malines-Brussels

I had to wonder what was going on at this point. Everyone had said until now that it required Cardinal Danneels' approval. Well, we had that. Now what? What was the hold-up? Another month passed, and we received the following letter from Archbishop O'Brien:

August 22, 2000

Dear General Callender:

Please excuse my delay in responding to your letter of some weeks ago. I have written to Cardinal Danneels but have not received a reply.

I am disposed, unless I hear otherwise from the Cardinal, to grant the request you asked. I will ask Msgr. Marceaux to make inquiry of interested individuals to ascertain the nature of the Mass desired, celebrant, frequency, etc. However, if it takes place, it should be carried out according to military and AMS expectations.

I hope this can be accomplished without too much time lapse or red tape.

Thank you for your patience and good will and do be assured of my prayers.

In the Lord,

✠ Edwin F. O'Brien
Archbishop for the Military Services

Well, what did that mean? We didn't know, but it could not be good. Not only had considerable time elapsed already, but now we had to wait for Fr. Marceaux to make an inquiry? Why didn't he just say, "go ahead"? That was the way they had handled the girl-altar-boy issue, and the variety of other radical, destructive liturgical changes over the years.

Vatican II was supposed to have rid us of clericalism; but apparently it was alive and well if anyone requested a Latin Mass.

This was no different than before. The excuse given for not allowing the Mass had been lack of the cardinal's blessing, up until he gave it. Now they were scrambling to find another excuse to block our request. Was Archbishop O'Brien going to let a local priest with a clear bias make the decision, instead of educating him that it was something to allow "generously"?

At that time, the headquarters of the U.S. Army in Europe (USAREUR) was located in Heidelburg, Germany, and it was commanded by General Montgomery Meigs, a four-star general. Soon we heard a rumor that USAREUR would not approve of our request because of a lack of funds. This was, of course, patently absurd. The same Army that buys $500 hammers can't afford a $50-a-week chaplain contract? Or, to be more precise, "another" contract chaplain, since they were already employed at many bases throughout Europe. I told Gen. Callender that I would happily pick up the cost myself, and that no funds were necessary.

"Never mind," he said. "If that's the best excuse they can come up with, this ought to be easy to fix."

Now that USAREUR was involved, however, the issue came to the attention of Brig. Gen. Callender's boss, Lt. Gen. Davis Weisman, who was the MILREP. Until now Lt. Gen. Weisman had pushed this issue off to his deputy, Gen. Callender, so that he could focus on the broader questions affecting America's role in the NATO Alliance. Suddenly, as the rumor had it, the question of a particular religious service in one of the many chapels on a military installation in Europe was being raised to the four-star level. And not by us.

The rumor was finally confirmed by an e-mail from the NSA Chaplain, Major Chester Egert, addressed to Gen. Callender:

> Yesterday (6 Sept, Wednesday) the USAREUR Chaplain, Ch (Col.) Haberek, sent an inquiry/guidance to my office through the 80th ASG Chaplain's office RE: the proposed Tridentine Mass at NSA. Bottom line: he's against it. He said (I paraphrase), "NSA/Brussels does not have his permission to conduct a Tridentine Mass."
>
> Let me preface the rest of this by saying that I think it would be nice to have a Catholic Mass at NSA, to draw the Catholics and Protestants together in this community. However, I have no authority or say in this matter. I'll support the decision either way. (I started collecting Catholic sacramental items for the sanctuary a year ago should we be directed by the Church to conduct Mass in the future.)
>
> The following I share with you, not because I have a side in the argument, but because Lt. Gen. Weisman will have to take the matter up with Lt. Gen. Jordon or Gen. Meigs in short order. Chaplain Haberek affirms that he will carry the matter to them if there's not a resolution at our level.
>
> First point: A commander, any commander or government official does not (repeat: does not) dictate to ecclesiastical authorities what they will and will not do. Doesn't matter if the ecclesiastical authority is Protestant, Catholic, Muslim or Buddhist. The government does not prescribe the activities of ecclesiastical bodies or their chaplains. Chaplains will be removed from a post before they violate their convictions or church policies.

Second: Permission from the Cardinal of Belgium and the Archbishop of the Military to conduct the Latin Mass is not the same as a mandate to conduct the Mass. The matter still remains with the senior chaplain authorities (in the Army). Chaplain (Col.) Haberek just happens to be (probably) the most senior Roman Catholic chaplain on active duty in the Army.

Third: The comparison of Fort Hood's support for Wiccan services with Latin Masses in Brussels is an apples and oranges comparison. What Wiccans do and Roman Catholics do doesn't even remotely relate to each other. The military doesn't prescribe to either group how they will conduct their meetings, unless the requested meetings are viewed to be contrary to the good order and discipline of the military.

Fourth: The mission of the military chaplaincy is to perform and provide for the religious needs of its members. Your role as a commander (Lt. Gen. Weisman's role) is to "support the free exercise of religion for all Army personnel." (AR 165-1, para 1-16, d) — not dictate the terms of surrender. The chaplaincy's role is "to hold religious services for members of the command to which they are assigned, when practicable." (AR 165-1, para 4-4) The USAREUR chaplain, for a number of reasons that are both ecclesiastical and practical, does not believe that this service is required on a military installation where Roman Catholic services are readily available in the local community in English and Latin. He does not wish to abridge Lt. Col. Sonnier's right to a Latin Mass. The Latin Mass is available in the local community. However, he does not have the requirement to provide it on U.S. government property any more than he's required to provide Southern Baptist services, Lutheran services, LDS, or any other.

When you argue that Wiccans meeting at Fort Hood justify Tridentine Catholics meeting in Brussels, he's going to ask you "Why then should we not provide 25 other distinctive services?" Your answer?

Finally, Chaplain Haberek is not "blowing smoke" when he says he'll go to the USAREUR commander on this issue. Is this the issue which you and Lt. Gen. Weisman wish to "fall on your sword" over?

The threatening tone of the email, particularly the last couple of lines, were rather irritating, but that was not the worst of it. There was no rationale here. There were not 25 different requests for distinctive services; there had been precisely two. One was for a protestant "Full Gospel Service" and for that, permission had been granted quickly and with much enthusiasm. The other was for a Latin Tridentine Mass, and it had resulted in more than a year's worth of lying, heel-dragging, backstabbing, bureaucratic obstacles, and now threats.

Suddenly they were arguing that "permission from the Cardinal of Belgium and the Archbishop of the Military to conduct the Latin Mass is not the same as a mandate."

It must have been irritating to Gen. Callender as well. He took advantage of a trip to USAREUR Headquarters in Germany to meet with Lt. Gen. Larry Jordan about the subject. Jordan was the deputy commanding general of the U.S. Army in Europe and the Seventh Army in Germany. In his capacity as the deputy, he was handling this question of a religious service so that his boss, Gen. Meigs, wouldn't have to. This was much like the routine way Weisman had previously turned the issue over to his deputy, Callender. At some point during this meeting, Callender was ordered by Jordan to back out and let it go.

I obtained a copy of the private email sent from Jordan to Weisman, who Callender and I both worked for. This email correspondence explains what happened perfectly well. Even though it is a private email from one general to another, it is important to understand the details of how the Tridentine liturgy was suppressed within the U.S. Military:

> from: dcg@cmdgrp.hqusareur.army.mil
> to: weismand@nato.brussels.army.mil
> cc: callender@nato.brussels.army.mil
>
> Dave,
>
> I am aware of a bubbling issue involving priest support for the Catholic community in Brussels. At the European School Council, Brig. Gen. Callender hit me with the subject. I had a brief conversation with him on the general issues and later obtained more extensive background and details from Steve Hayward and the USAREUR Chaplain. My intent at this time is to clarify the USAREUR position and avoid any further misunderstanding.

It appears to me that Earnie Callender may have misinterpreted my willingness to receive and objectively consider the request to hire a contract priest, and my acknowledgment of his report that the Cardinal of Brussels and Archbishop of the Services (O'Brien) did not object, as an indication of my support or tentative approval of the request. Such was not the case. I could not, and would not make a decision based on that brief conversation and general outline which lacked complete details.

Several facts weigh on this issue:

Catholic priests are significantly in short supply across the Army and the command (and indeed, the country).

The reported number of Army and USAF Catholic personnel assigned to Brussels is reported as 35.

Catholic mass in English is available in at least two locations in/about Brussels (1.5 and 10 miles, respectively). Latin mass is available in at least four locations (closest is 4 miles).

USAREUR QOL (Quality of Life) standard is 30 min travel time for religious services. However, several places across the command have many more Catholic personnel, are more distant from available services, but have neither military nor civilian priests available.

While the Cardinal of Brussels and Archbishop O'Brien may not object to hiring a contract priest from Italy in this specific circumstance, it is the Chief of Chaplains who is responsible for Army religious programs, and U.S. Army Europe who is responsible for resourcing religious programs here.

Brig. Gen. Callender alleged that past disagreements between the USAREUR Chaplain and one of the field grade officers in Brussels requesting this support may have contributed to the current impasse. Even if that were true, which is not certain to me, the point would be moot. I have not based my decisions on input from any single source.

I am not prepared to support this request. Adequate religious support appears readily available (with both English and Latin mass offered). The request is not supportable based on the USAREUR QOL standard; the number of assigned personnel does not justify the hiring; and it would be impossible to justify given the level of priest support across the command and in other specific locations where identical or worse levels of support are available. I see no rationale for an exception to policy, and the extra expenditure would deny needed support elsewhere.

If I have not stated the facts correctly, or if additional relevant information is available, I will gladly consider it.

Regards, Larry

Lt. Gen. Larry R. Jordan
Deputy Commanding General
United States Army, Europe

After General Callender read the email above to me, I could hardly contain my anger. "Sir, where is he getting this? Can we tell him he has some erroneous information? Who told him we're trying to hire a priest from Italy? What 'extra expenditure' is he talking about?"

General Callender wasn't his usual animated self. He had been frustrated from day one in his dealings with USAREUR on a variety of other, unrelated issues.

Why did Jordan think that we were trying to hire a priest from Italy? A recently ordained, English-speaking priest of the Institute of Christ the King had just finished his studies at Gricigliano and had been assigned to Belgium, and Cardinal Danneels had given consent to letting him offer our Mass. That was the only Italy connection we knew of.

Sure, there were Latin Masses in the area, but the homily was always in French or Flemish. That was fine for us, in the Sonnier family, but what about those Americans who didn't speak French or Flemish? And we knew priests were in short supply everywhere. That was precisely why I was trying to get the Army to become open to accepting priests from traditional orders.

"Sir, for Pete's sake, tell him it won't cost a dime."

"I've been ordered to stay out of it."

"Who do I go to now? Should I write to him myself and tell him I'll cover any costs?"

"I'm sorry, Dave. I've been called down on this one. I did my best."

Over the next few days, I made some unsuccessful attempts to convey through the chain of command some of the problems with General Jordan's email. In one such attempt, Callendar's JAG, Col. Klein, contacted General Jordan's chaplain, the senior chaplain for the USAREUR Fr. (Col.) Jerome Haberek, a Catholic chaplain. Klein pointed out the problems with the case General Jordan was trying to make. Somewhere in the conversation he was interrupted:

"I can stop this Tridentine Mass from happening!" Haberek said angrily.

"Why would you want to do that?" Klein asked.

"You just watch. I'll stop it."

It really was fruitless. Jordan had spoken, and that was the end of it. It didn't matter what the facts were. He was going to follow the advice of his chaplains, and Fr. Haberek didn't want it to happen, Fr. Marceaux didn't want it to happen, and they didn't care what Cardinal Danneels and Archbishop O'Brien had to say. They definitely didn't care what the Holy Father had to say. They were going to ensure that there was no Tridentine Mass, ever, in one of "their" chapels. They would, of course, permit witchcraft, but not a Tridentine Mass. On July 9, 2000, for example, the featured headline in *Stars and Stripes* had been "Pagans Celebrate Diversity, Togetherness," featuring photos from the Mannheim military Pagan coven at the Taylor Barracks Chapel in Mannheim, Germany. "In February 1998, the Pagans in Mannheim took control of the keys to the chapel at Taylor Barracks."[1] This could not have been done without the full consent of the Catholic clergy in the chaplains' corps. They, Catholic priests, were willing to turn over a base chapel to Wiccans, but they would go as far as to lie to prevent a Tridentine Mass.

Now, with more than nineteen years in the army, a possible retirement and reassignment less than a year away, I knew what I had to do.

---

[1] *Stars and Stripes*, Sunday, July 9, 2000, p. 5.

Pagans celebrate diversity, togetherness

BY ERIC B. PILGRIM
Stars and Stripes

Tammy Anderson meditated for a second and then plunged the knife into the bowl.

Smoke rose from burning incense, danced around her, seemed to envelope her. Other worshipers sat in a circle, tapping on drums, humming in rhythm to their own beats. Anderson scratched a five-pointed star, then a circle, into the bowl of grain, sat back down atop her own drum and lead the others in an earthy, old Pagan chant that resonated from the concrete walls of the military chapel.

As the Taylor Barracks Chapel bells declared the third annual Pagan Mini-Conference in Mannheim, Germany, on Saturday, about 80 military and nonmilitary Pagans and Wiccans from around Germany celebrated their diversity and togetherness with dance, chants and instruction.

Armed with candles, incense, stones, chants and friendship, the only problem they encountered Saturday was the threat of rain. Many of those attending welcomed the rain, but admitted that it was causing some logistical problems with planned activities outside.

Some of those attending say their religious dealings with others don't always go so smoothly.

Lisa Mikles, a previous leader of the Mannheim coven, said Pagans and Wiccans are often misunderstood.

"A lot of people think we are

It's that we see the divine in nature."

Mikles said this misunderstanding can lead to problems.

"A lot of times we are met with opposition," Mikles said. "Let's face it, we live in a world where many people are Christians. And if they aren't Christians, then they are Muslims or Judaic."

Many Pagans consider their religion to be the oldest of all because it rose from the earth. Mikles said that is one reason why she believes their numbers are increasing. One Army handbook estimates there are 50,000 Pagans in the U.S. military. Mikles believes there are 4,000 in the U.S. military in Europe.

But both of those figures are hard to determine because of the nature of the religion itself. There are many different autonomous covens or forms of belief within the Pagan faith.

In fact, according to literature handed out at the conference, "it is rare to find two modern Pagans that have the same set of beliefs or even define their religion with the same terms. There are some basic beliefs that tie Pagans together, but no one set of rules or even of definitions."

PaganPaganIn spite of that lack of standardized doctrine — or maybe because of it — membership in Pagan organizations appears to be growing.

"People have lost their connection with the divine and the earth itself," Mikles said. "We're now seeing people come back to this."

Mikles said some of their

PILGRIM/*Stars and Stripes*

Tammy Anderson, leader of the Mannheim military Pagan coven, stabs a bowl of grain with a knife as part of a ritual to honor sacredness and oneness during Saturday's conference at Taylor Barracks Chapel in Mannheim.

them by providing places to worship. The Pagan religion is recognized by the U.S. military. In February 1998, the Pagans in Mannheim took control of the keys to the chapel at Taylor Barracks. Mikles said the group has seen a lot of growth since that time.

ginning to learn they have rights," Mikles said. "So a lot more people are starting to come out of the broom closet, so to speak."

But despite several people coming out of that closet in recent years, many still prefer to cloak their beliefs in privacy. Pagan

ing and misinformation out there," said Amy MacWatt, the wife of Ian Stewart MacWatt who founded the coven in Mannheim. "Take witches for example. What is a witch? She is a woman of wisdom who has developed years of knowledge. But they were burned at the stake

CHAPTER 18

MEET THE PRESS

BEGAN TRYING TO CONTACT SENATOR TRENT Lott's office immediately. At that time Senator Lott was the Senate Majority Leader, but he was also my home state senator. Any serviceman has a right to bring a grievance to his congressman or senator. Furthermore, he had nominated me to West Point about 24 years earlier when he was a Mississippi congressman.

I was put in contact with one of his assistants. I explained my complaint, and faxed a copy of the documents related to it. I also sent a copy of a letter I would be mailing to Senator Lott.

"You'll hear back from someone within 2-3 months," he said.

"That's too long," I said. "Time's running out on me here and I'm going to end up getting reassigned within a year."

He paused. "Well, if you really want to do something about it right away, I'd pass it to the press."

What did I have to lose at this point? Why not?

"Call Rowan Scarborough," he said. "He's a reporter for the *Washington Times*. He's got a way of turning things around in a situation like this. He'll make sure it meets the public eye in the right light. But I've got to warn you, there are sometimes repercussions for going to the press, even if you're right."

By now, I could care less about repercussions. I didn't care if they stood me up in front of a firing squad and shot me. The way I saw it, I was serving in an Army in which the Catholic chaplains had turned against their legitimate Church authority. If my insignificant military career became a casualty, I would be glad to see it go.

In a day or two Scarborough got back with me, and he was quite efficient. I think he was writing the article as he spoke with me. He wanted me to fax a couple of other documents such as the email from Lt. Gen. Jordan to Lt. Gen. Weisman, and the letter from Archbishop O'Brien.

Scarborough was a genuine professional. He was very careful with the timing when he submitted it, knowing that if it hit the papers on a weekday rather than on a weekend it would end up in the *Early Bird*, a condensed summary of all the day's articles which circulates throughout the Department of Defense every morning.

The following article appeared on page 7 of the *Washington Times*, October 10, 2000, as well as in the widely-read *Early Bird*:

Catholic Denied Mass in Army Chapel

Rowan Scarborough

The U.S. Army in Europe will not allow traditional Catholics to use a base chapel, although the military has opened similar facilities around the world for witchcraft and pagan rituals.

The officer seeking to use the chapel blames the denial on a "politically correct Army."

But an Army spokesman says the branch's European command does not want to pay the cost of hiring a priest when personnel can go off base to churches where Mass is conducted in Latin. Catholics have the option of attending Mass that is celebrated in the vernacular, meaning the local language, or in Latin. The vernacular is the most common choice.

Lt. Col. David L. Sonnier, an information management officer, said in an interview that his request was turned down by Lt. Gen. Larry R. Jordan, deputy Army European commander. Col. Sonnier serves at the NATO Support Activity, a small Army base in Brussels that provides logistics and personnel support for the U.S. delegation to NATO headquarters in Brussels.

The colonel says the chapel hosts two Protestant services weekly. He said he was told that Gen. Jordan refused the Catholics' request because he feared a multitude of other religious groups then would want chapel time.

"Just let us have access to it," said Col. Sonnier, estimating that 20 Catholics want the service. "All we want is the Army to allow us to use chapels like the Wiccans and pagans. This is politically correct to an extreme when pagans and Wiccans are allowed into a chapel and traditional Catholics are blocked entry."

The Army says it's not a question of access, but of cost. The pagan sects supply their own ministers. "In order to (have a Catholic priest say a Latin Mass), the Army would have to contract out to a private priest," said Col. Carl Kropf, a spokesman in Heidelberg, Germany. No Catholic Army chaplain was available.

Col. Kropf said Gen. Jordan "didn't deny the use of the chapel, which someone may have alleged. But he wasn't going to let the contract [*sic*] for something that is available elsewhere. It's a resource issue."

The colonel [Kropf] denied that Gen. Jordan is concerned that many other religious groups subsequently will want chapel space. "I think this is a situation where a service is available in the local community that serves the needs of the people."

Col. Kropf said the Army suffers "a tremendous shortage of Catholic chaplains in particular."

On the cost issue, Col. Sonnier said: "They don't have to pay. We made that clear. That's not an excuse. If the Army doesn't pay the cost for a priest, I will. Also, other members of the community. What is their next excuse?"

He estimated the weekly cost at $120.

Col. Sonnier, a 19-year officer who is married and has five children, has written to Senate Majority Leader Trent Lott, Mississippi Republican, seeking his help. Col. Sonnier, a Mississippi native, was nominated to West Point by Mr. Lott.

"One has to wonder whether the same standards that were applied for the pagans and Wiccans are being applied in our case," the colonel wrote, adding: "It appears that serious, orthodox Catholics in good standing within their denominations are now being denied access to the chapels by military leaders who supposedly want to avoid having too many different religious services.

"I am quite certain that our fellow Mississippians will be interested in keeping their children away from the recruiting stations until our morally disoriented military undergoes some serious reform."

The colonel says he gained approval for Latin chapel services from Archbishop Edwin F. O'Brien of the U.S. Military Archdiocese, and Cardinal Godfried Danneels of the Archdiocese of Mechelen-Brussels. Permission from the Catholic hierarchy is required before Latin Masses may be celebrated rather than the more common vernacular Mass.

As Col. Kropf pointed out, "The Catholic service for people in the U.S. and Europe is an English-based service. That's what you get when you are in Fairfax, Virginia."

Archbishop O'Brien's Washington office said he was traveling and unavailable for comment.

Defense Department regulations urge commanders to accommodate different religious groups: "Commanders should provide for the faith needs of all service members and accommodate their requests to conduct religious observances on military installations when these observances will not adversely impact military readiness, unit cohesion, or standards of good order and discipline or violate health or safety standards."

A July 9 article in *Stars and Stripes*, a newspaper for military people, said Army pagans in Mannheim, Germany, have the keys to a base chapel where they hold regular gatherings.

Col. Kropf said the chapel in question is excess property, and there is no extra cost in providing the building.

The military's tolerance of Wiccans and pagans became national news in 1999, when it was reported that the Army base at Fort Hood, Texas, the largest U.S. base, provided a Wiccan coven a campsite to hold regular ceremonies.

The military's 20-year-old chaplain handbook recognizes more than 200 religious faiths, including Wiccans, the Church of Satan and Rastafarians.

Col. Sonnier said his research shows that at least 12 military bases worldwide host pagan or Wiccan ceremonies.

Soon there were other articles. ZENIT's version, dated October 11, and largely a resumé of the *Times* coverage quoted above, appeared in various locations: "The New Army: Pagans Can Use Chapels but Latin Mass Is Forbidden."

Ohio Congressman James Traficant (D-Ohio) took the issue to the House floor.[1] He called out the Army in one of his typically colorful fulminations:

If that's not enough to shred the Bible. The Army does allow and permits witchcraft and pagan ceremonies at the base ... unbelievable. It's time to call in the dogs,

---

[1] "Ohio Congressman Blasts Army for Denying Latin Mass in Chapel," CNS Senior Staff Writer Jim Burns, October 11, 2000 (not online).

> throw the coffee grinds on the fire, the hunting's over.
> When the U.S. Army allows Satan in one door and will
> not allow God in the other door, America is so screwed
> up, we don't know where we're going. An America with-
> out God is an America that the founders never planned.

To those who would contend that taking this issue to the media was going too far, I would respond with a reminder that the issue had already been elevated to the level of four-star general officers and already been carried to the office in the Vatican responsible for handling such issues. Sometimes the clarification of an issue is only possible through public exposure, and in these articles, we can see that some clarification was finally taking place. While the bureaucrats in both the U.S. Military and the Military Archdiocese were focused on using the convenient, if ridiculous, excuse that "it would cost too much," they had to choose the words for their public responses very carefully. These public articles never repeated the false narrative that "Vatican II changed the Mass to the vernacular." They could not say that because it wasn't true. We also do not find the word "schismatic" associated with the words "Latin Mass" in these articles. Until now these two strings of characters had almost always gone together. Instead, we find in *The Washington Times* article that "Catholics have the option of attending Mass that is celebrated in the vernacular, meaning the local language, or in Latin. The vernacular is the most common choice." Although this statement was mostly false at the time, because most Catholics really couldn't "choose" in the year 2000, it was a change from the media narrative that had been in force since the Second Vatican Council. I'm quite sure that for some people this was the first time that they had ever heard that they had a choice. If anyone reading one of these articles subsequently attempted to exercise the option to attend a Latin Mass, there's a high probability that they could not find one within a reasonable distance.

I also wrote to His Eminence Dario Cardinal Castrillón-Hoyos in his capacity as Prefect at the Pontifical Commission *Ecclesia Dei*. He had recently replaced Cardinal Felici in that position. If he had the moral and physical courage to sneak into the residence of drug kingpin Pablo Escobar disguised as the milkman, certainly he would have to courage to assist us in this case.

CHAPTER 19

EXCUSES, EXCUSES

OON AFTER THE FLURRY OF ARTICLES, I WAS informed that I had never submitted a formal request to use the NSA chapel. I was directed to submit a request under the provisions of a "distinctive faith group service." I did so, even though I didn't see the need for it. It seemed to be another hoop to jump through to make this Latin Mass finally happen. By now, I knew the strategy. This request would be submitted and then there would be nothing done for an incredibly long period of time, as I sat and waited for some response. I submitted the request:

> Subject: Request for use of the NSA Chapel Facilities
>
> 1. I request use of the NSA Chapel facility for a weekly traditional Mass, according to the 1962 *Missale Romanum* under the provisions of *Ecclesia Dei*, as authorized by God-fried Cardinal Danneels and Archbishop Edwin F. O'Brien. I understand that this "distinctive faith group service" would need to be held or contracted on an exception to policy basis because the Army is unable to provide a military chaplain, DOD civilian, or volunteer to meet this request. This service will be attended by 20+ service members and their family members. I understand that this exception must be approved by the USAREUR Chaplain.
>
> 2. I also understand that this service must be sponsored and supervised by an assigned chaplain. This service and the offerings that may be received will require the support of Army personnel to fulfill the requirements of AR 165-1.
>
> 3. P.O.C. for this request is Lt. Col. David L. Sonnier, U.S. Delegation IMO.

On the morning of November 3, 2000, Col. Jerome Haberek, the USAREUR Chaplain, arrived from headquarters to give a verbal response to our request. He had been one of the priests assigned to Fort Bragg who had refused to speak with me about our petition there.

Upon his arrival he first met in private with Lt. Gen. Weisman
and Brig. Gen. Callender. I was not allowed to attend this meeting,
so I have no idea what he told them. More importantly, I was
not allowed to defend our case. I won't know what he told them
until Judgment Day. I'm looking forward to finding out then, but
it will not matter at that point.

Meanwhile, I gathered some friends, colleagues, and associ-
ates; some interested in attending the Latin Mass, others just to
be witnesses to the ensuing public meeting. One of the people
who came to the meeting was a Missouri Synod Lutheran who
seemed to be finding it difficult to believe that we were not
allowed to use the chapel. He had to come and see what the
big deal was for himself.

Haberek clung doggedly to his position that we would not
be allowed to use the NSA Chapel for the traditional Mass but
rather should attend Mass in a local parish. He nervously read
from a prepared statement. I was able to obtain a copy of it so
that I could more closely examine the rationale used to prevent
the Latin Mass on this military installation:

> The availability of Catholic services in the local com-
> munity is a necessary part of our identifying need. We
> accommodate religious support heavily dependent on
> whether the legitimate needs can already be satisfied
> or not, either on post or off post. We use this rule for
> all faith groups. In this case because of a shortage of
> military priests, we do not have a Catholic mass supplied
> at NSA chapel. Our requirement and desire is either to
> perform the services for our people (when a military
> chaplain of the appropriate faith group is available)
> or to provide a service when we don't have the assets
> and/or when the fewness of people do not justify an
> entirely new service. In this case, the fact that Catholic
> needs are totally satisfied in the community (where the
> majority lives) answers our requirements and does not
> force us to make all sorts of duplicate expenditures or
> administrative work.
>
> Thus in this case (as in all others), the two English
> masses convenient to our people and the Latin Mass, up
> until now have fully satisfied our requirements. If we were
> to discover through a needs assessment, that we needed
> a Catholic Mass at NSA we would most likely have to fig-
> ure out a way to provide an English Mass there to serve

> the needs of the vast majority. That English Mass, after all, would have to satisfy completely the requirements and preferences of the Roman Catholic Church and fully responds to the obligations of English-speaking Catholics.
>
> We are not in a position (nor could we ever be) to attempt to satisfy the personal liturgical preferences of every individual or small group of people within any religious tradition. We use whatever flexibility we reasonably have (including providing off-post civilian services when they are unique), but we can't create a policy to supply all at all times. In fact, in conjunction with the advice of the civilian church endorsers (CMAF), our policy and ministry is executed within the realm of the practical.
>
> In this case, the off-post option is perfect, language is not a problem since Latin is used. Individuals can find their particular brand of liturgy as well as strengthen and support the existing small communities. The Catholic Church's preference for non-proliferation of unnecessary liturgies (especially Latin ones) when very few people are involved is satisfied. It seems a win-win as long as spiritual welfare of all is the criterion.

After reading his statement aloud, he asked us for our questions on the issue.

I asked him to repeat the line "... Catholic needs are totally satisfied in the community..."

Referring to the local English-language parish, I asked: "Are you aware that the parish does not have First Confession before First Communion, in violation of the Code of Canon Law? There are other serious problems as well..."

Before I could finish, he snapped nervously: "I don't want to get into difficulties you're having with your parish."

Referring to his text: "Were we to discover, through a needs assessment, that we needed a Catholic Mass at NSA we would most likely have to figure out a way to provide an English Mass there to serve the needs of the vast majority."

I pointed out: "We are not requesting an English Mass."

"The vernacular is the standard," he responded.

"Can you give us a source for that statement?" I asked.

He could not.

"Call Phil Hill at the Office of the Chief of Chaplains," he said. Phil Hill, we found out later, was a priest.

Lt. Col. Klein said, "that's in contradiction with Canon 928, which states that 'Latin is the typical language of the Latin rite...'"

"We don't want to get into canon law," he said curtly.

I then questioned the statement that "we are not in a position (nor could we ever be) to attempt to satisfy the personal liturgical preferences of every individual or small group of people within any religious tradition." I said we were not asking anyone to do that.

Referring to his statement that "the Catholic Church's preference [is] for non-proliferation of unnecessary liturgies (especially Latin ones) when very few people are involved," I asked him "What is your source for this statement that the Church prefers a non-proliferation of 'unnecessary' liturgies? It is my understanding that the value of a single Mass, offered reverently, is infinite."

There was no answer.

At the end of the meeting, it was clear that he had no legitimate arguments, and he could offer no clear reasons for his decision. It was all based on personal bias. Fr. Haberek concluded this meeting in a room full of Catholics by turning to the Protestant chaplain and asking him lead a closing prayer.

After the meeting I immediately contacted Archbishop O'Brien by phone to inform him of the situation. The first question I asked was whether we had permission to proceed with the Mass. He affirmed that we did, but then when I informed him of Fr. Haberek's objections, he said "I was told that it was the command that was opposed to it." I assured him that the command did not care one way or another, and that in fact it was Fr. Haberek and Fr. Marceaux who were opposed. He didn't have time to discuss the issue over the phone, so I sent him the following letter by fax. In the cover letter I enquired as to whether we could proceed despite Fr. Haberek's objections.

3 November 2000

MEMORANDUM FOR: Archbishop Edwin F. O'Brien
Archdiocese for the Military Services

SUBJECT: Request for Permission to Proceed with Traditional Mass

1. A meeting was held on 3 November 2000 with the USAREUR Chaplain, who denied permission to proceed with the traditional Mass at the NSA Chapel, but offered no substantial reason for doing so.

2. At this point, we are unable to proceed due to what appears to be the personal bias of the USAREUR Chaplain. No commander at any level of command presents any objections at all, since no resources are requested other than the space of the chapel. This initiative has the approval of the NSA Commander, and apparently the 80th ASG Commander. The USAREUR Deputy Commander stated publicly (through his PAO) that he is not stopping us from using the chapel.

3. Request your ecclesiastical permission to proceed with the traditional Mass on a trial basis. This permission will be for Fr. William Hudson, Institute of Christ the King, who has been invited to Belgium by Cardinal Danneels for this purpose.

4. POC this request is Lt. Col. David L. Sonnier

Information Management Officer

I immediately sent an email to Lt. Gen. Weisman and Lt. Col. Don Isbell, with a courtesy copy to the NSA Chaplain:

Sir:

I just got off the telephone with Archbishop O'Brien, and he said that we have ecclesiastical approval to go ahead and use the chapel. He agreed that there were no reasonable arguments presented in the meeting today with Col. Haberek, and he insisted that he was being told that it was only "the command" that was presenting an obstacle. I don't know who is telling him this, since no commander at any level has any objections.

He asked me to go ahead and fax a request directly to him and he will respond to make it formal. Lt. Col. Klein came in to hear the last part of the conversation.

I sent the fax, along with a recommendation that he contact the Army Chief of Chaplains and let him know that he should let us proceed despite Col. Haberek's objections.

Lt. Col. Sonnier

The response from Archbishop O'Brien came in the form of a letter sent by fax:

Dear Colonel Sonnier:

Thank you for your fax, just received.

I am not giving anyone "permission to go ahead despite Colonel Haberek's objections." Nor was that the terminology used by either of us in our telephone conversation this morning.

I have called Monsignor Hill in Ch. Gunhus' office to let him know the nature of my permission:

Since there is no objection from the local diocesan bishop, Cardinal Danneels; and since there is no evident and convincing pastoral reason to forbid the traditional Mass; I am willing to grant permission for such a Mass on a trial basis, the nature of which is to be addressed.

In the Lord,

✝ Edwin F. O'Brien
Archbishop for the Military Services
Cc: Msgr. Hill

A few days later, on November 6, I received the following email from the NSA Chaplain, Chester Egert, in response to the email I had sent to Weisman and Isbell on November 3:

Sir:

Thanks for copy of the email. Perhaps LTG Weisman told you, but in case he didn't, let me. General Meigs is about to distribute policy guidance in support of Chaplain (COL) Haberek's position. With such guidance, it keeps the decision out of our hands and the answer is still "No service."

I don't have the freedom to ignore that counsel, and unless LTG Weisman tells me to do it, my hands are tied. Presently, my Chief of Chaplains has told me "No service." He has command authority over my activities.

Archbishop O'Brien is a player in this matter, but not the only player. I can proceed when Chaplain Haberek, my Chief of Chaplains, or LTG Weisman tells me to. As of Friday's meeting, LTG Weisman is in unison with General Meigs.

Now it became clear. Fr. Haberek, a Catholic priest, was telling Archbishop O'Brien that it was "the command" that didn't want the Latin Mass. But what was he telling the command to convince them that we somehow had to be treated differently from any other group of soldiers seeking a religious service? We had not been allowed into that discussion. Now we knew exactly where the problem was.

I received a response from His Eminence Dario Cardinal Castrillón-Hoyos to the letter that I had sent to him on October 2. It was rather disappointing to read that "we must respect the will of the Holy Father with regard to the present provisions for celebrations according to the 1962 Roman Missal as he has manifested it." That didn't take much courage! Why didn't he clarify what the "present provisions" were? He no longer sounded like the apostolic successor with legendary courage who would risk martyrdom by sneaking into the residence of a drug kingpin.

PONTIFICIA COMMISSIO
« ECCLESIA DEI »

N. 9/95

Rome, 31 October 2000

LTC David L. Sonnier
Memlingdreef, 62
B-3090 Overijse
BELGIUM

Dear Lieutenant Colonel Sonnier,

I wish to acknowledge your letter of 2 October 2000 in which you detail the many difficulties which you have encountered and continue to encounter in seeking to have the celebration of the Mass according to the 1962 Roman Missal in the military chapel at the United States Army Base in Brussels. I sincerely regret the unfortunate bureaucratic obstacles which presently prevent this celebration from taking place. Nevertheless, we must respect the will of the Holy Father with regard to the present provisions for celebrations according to the 1962 Roman Missal as he has manifested it.

As you have already overcome so many obstacles, let us trust that with the grace of God and the intercession of His Holy Mother in due course you will also overcome the present ones.

Commending myself and the Pontifical Commission "Ecclesia Dei" to your prayers and assuring you of my blessing, I remain

Sincerely yours in Christ,

Darío Cardinal Castrillón Hoyos
President

The disappointment of what appeared to be a display of weakness hit my wife harder than it hit me. In retrospect, however, what else could His Eminence do in his position? A strongly worded letter to Archbishop O'Brien or Cardinal Danneels would have been pointless; *they* were not the problem. The problem was that Catholic chaplains had rejected the authority of the pope by rejecting these "present provisions," and they had maneuvered to convince the senior officers that the people making this petition were to be treated as an exception.

Soon I received a letter from Senator Lott, along with a letter from Major General Lennox, Chief of Legislative Liaison:

> Dear Lieutenant Colonel Sonnier:
>
> Thank you for writing to me to request my assistance in helping you to get authorization to conduct traditional Mass services at the NATO Support Activity (NSA) Chapel in Brussels. Attached is a letter from the Army which responds to my inquiry on this matter.
>
>> U.S. Army commanders are responsible for providing religious, spiritual, moral and ethical support to authorized personnel in their command. In this light, the Deputy Commander of U.S. Army Europe and Seventh Army (USAREUR), Lieutenant General Larry R. Jordan, has determined that your religious needs are already being satisfied through existing services within the community. He notes that Latin Mass is conducted at four different parishes within a 15-mile radius of NSA which complies with the Quality of Life Standard for having religious services available within a 35-minute drive.
>
> Please review the Army's response and determine whether this response is satisfactory. Thank you again for affording me the opportunity to assist you with this matter. With very best wishes, I am
>
> Sincerely yours,
>
> Trent Lott

Enclosed was a letter to Senator Lott from Maj. Gen. Lennox, Chief of Legislative Liaison, stating the same thing. Soon after this I also received a copy of a new policy letter from the Commanding General, Headquarters of the United States Army, Europe, and Seventh Army:

DEPARTMENT OF THE ARMY
HEADQUARTERS, UNITED STATES ARMY, EUROPE, AND SEVENTH ARMY
THE COMMANDING GENERAL
UNIT 29351
APO AE 09014

AEACH-ZA (165)   18 December 2000

MEMORANDUM FOR SEE DISTRIBUTION

SUBJECT: USAREUR Command Policy Letter 28, Accommodating Religious Needs

1. Reference AR 165-1, 27 February 1998, Chaplain Activities in the United States Army.

2. Chaplains are responsible for accommodating the religious needs of all personnel, regardless of their religious affiliation. Each individual must be provided the opportunity to practice his or her religious beliefs freely.

3. Chaplains need not establish worship services to accommodate each religious affiliation. They may, however, help individuals grow through prayer and scriptural study, and provide literature that supports the individual's faith or denomination. Distinctive faith group leaders (DFGLs) also may provide ministry support as an exception to policy if a military chaplain is not available to meet the requirements of a distinctive faith group. Requests for distinctive faith group support should originate from soldiers.

   a. AR 165-1 prescribes policy on DFGL certification, DFGL use of chapels, and guidelines for conducting distinctive faith group services. To request certification, DFGLs will send the information specified in AR 165-1, paragraph 5-5, to the area support group (ASG) chaplain. The ASG chaplain will review the request, prepare a recommendation, and forward both to the USAREUR Chaplain, the approval authority for DFGL certification in USAREUR.

   b. DFGLs are authorized to conduct worship services, but may not establish a church, parish, congregation, or mission. DFGLs are not authorized to provide a full range of programs or activities normally expected in a civilian parish, church, or mission for a particular denomination or faith group (for example, youth ministry, religious education); nor are DFGLs authorized to assume control or management of military facilities.

4. Commanders usually can accommodate the religious needs of personnel in their command. If resources restrict the religious support that can be provided, accommodation may be limited to providing the essential elements of religious services required by a distinctive faith group, or by offering a broad or collective service that combines the needs of several distinctive faith groups.

MONTGOMERY C. MEIGS
General, USA
Commanding

DISTRIBUTION:
A (UPUBS)

This letter is available at http://www.aeaim.hqusareur.army.mil/library/home.htm.

18 December 2000

USAREUR Command Policy Letter 28, Accommodating Religious Needs

1. Reference AR 165-1, 27 February 1998, Chaplain Activities in the United States Army.

2. Chaplains are responsible for accommodating the religious needs of all personnel, regardless of their religious affiliation. Each individual must be provided the opportunity to practice his or her religious beliefs freely.

3. Chaplains need not establish worship services to accommodate each religious affiliation. They may, however, help individuals grow through prayer and scriptural study, and provide literature that supports the individual's faith or denomination. Distinctive faith group leaders (DFGLs) also may provide ministry support as an exception to policy if a military chaplain is not available to meet the requirements of a distinctive faith group. Requests for distinctive faith group support should originate from soldiers.

a. AR 165-1 prescribes policy on DFGL certification, DFGL use of chapels, and guidelines for conducting distinctive faith group services. To request certification, DFGLs will send the information specified in AR 165-1, paragraph 5-5, to the area support group (ASG) chaplain. The ASG chaplain will review the request, prepare a recommendation, and forward both to the USAREUR Chaplain, the approval authority for DFGL certification in USAREUR.

b. DFGLs are authorized to conduct worship services, but may not establish a church, parish, congregation, or mission. DFGLs are not authorized to provide a full range of programs or activities normally expected in a civilian parish, church, or mission for a particular denomination or faith group (for example, youth ministry, religious education); nor are DFGLs authorized to assume control or management of military facilities.

4. Commanders usually can accommodate the religious needs of personnel in their command. If resources restrict the religious support that can be provided, accommodation may be limited to providing the essential elements of religious services required by a distinctive faith group, or by offering a broad or collective service that combines the needs for several distinctive faith groups.

Montgomery C. Meigs
General, US Army
Commanding

Unbelievable. This policy letter had apparently been generated because of the stir I had created. But what was I doing that was so unusual? This was a request to worship as Christians had for over a millennium and a half up until the mid-1960s. This was a request to be allowed to worship in a military chapel in a rite that was supported by the highest authority in the Church.

Yet to these chaplains and senior officers involved, it was more distasteful than a request for a Wiccan service.

At this point I was starting to get email and phone calls from the Special Forces branch field grade assignment officer. They were going to reassign me in a few months, and there was no position in Brussels for a Special Forces lieutenant colonel. But I saw no reason to stop pushing the issue. If I could hold out long enough we would eventually succeed.

There was a possibility for me to take a command position at my rank. Don Isbell would be leaving within a year, and I was as qualified as anyone else to be his replacement. That would allow me to stay on in Brussels and wrestle this problem to its logical conclusion. They couldn't keep traditional Catholics out of Army chapels forever.

Don and I discussed the idea. "Are you sure you can convince General Weisman to agree to this?" I asked him.

"I think I can convince him."

We agreed upon a timetable for change of command, finding a replacement for myself as the Information Management Officer, and even a two-week vacation for both of us.

"You're nuts!" Lorri shouted, when I told her about the plan.

"You don't want to stay in Brussels and see this through to the finish?" I asked, genuinely surprised.

"Absolutely not! This stupid Army has gone to hell in a hand-basket. You tell me what's so hard about getting a Latin Mass in that chapel? They don't want traditional Catholics in their Army! They don't want people like you in this Army!"

Whew! This drawn-out affair had obviously taken its toll on my family. I knew I wouldn't be able to continue in the Army without her support.

"What do you want me to do?"

"I want you out. I want you out by this summer! I think God wants you out. If God wanted us to have a Latin Mass in the Army, He would have made it happen by now. He would have turned the hearts of some of these Novus Ordo priests. I think He wants you out, because our country has turned against Him. And just by being in the Army you're defending a country that has turned against God. Why serve in an Army that forbids the Latin Mass, promotes Wicca, and defends a nation that has turned against God? Why?"

After a glass of wine, a long talk, some flowers, and a little extra attention, I convinced her that I should stay in the Army a little longer and try to make the best of the situation. Meanwhile, I would continue to work on the Brussels problem until my reassignment the following summer.

But reassignment for what? The more I thought and prayed about it the more I came to the same conclusion: there was no reason for me to be in the Army. Lorri was right. If they couldn't do something as simple as allow the Latin Mass for those of us with (to use the pope's own words) "rightful aspirations" to have it, then there was no sense in continuing to be a part of the organization. It was a waste of time. Defend what? A corrupt America, whose corrupt president was fondling women in the Oval Office? An America whose corrupt Catholic clergy could pull the kind of stunts I witnessed in the last six years and get away with it?

No, if I was going to stay in, I would remain in Brussels and see it through to the finish. If that didn't work out, I would leave the service. Don and I presented the plan to Gen. Weisman. He came by my office later to give me his decision:

"Dave, I can't have you take command of the NATO Support Activity. First, you don't have your children in the DODDS[1] school, so you're not supporting our program there, and second, you're in the middle of this public pissing contest with USAREUR and a bunch of chaplains. I just can't do it."

That meant we would not be staying in Belgium much longer; one way or another we would be leaving.

It dawned on me that there had been no response to our October 19 request to use the chapel as a "distinctive faith group." Certainly, under this new policy there could be no denial. I began prodding for a response. But the denial of our request came anyway. The rationale? "This service is available within two miles of the installation and is able to accommodate the religious needs of those requesting this service." In other words, since the local ordinary allowed the Latin Mass, we didn't need it.

So, at Fort Bragg, where we had a desperate need for relief from the liturgical atrocities, the Latin Mass was not allowed

---

[1] DODDS was the "Department of Defense Dependents Schools" (which changed names around this time). There was a school in Brussels for the children of service personnel assigned there, but we had our children attending *L'Institut Saints Pierre et Paul*, where they received a genuine Catholic education.

because the local ordinary didn't allow it. Now, the Latin Mass was not allowed because the local ordinary *did* allow it. Either way, regardless of what the local ordinary was doing, the Latin Mass would not be allowed in a military chapel.

Who was left that I could appeal to?

Archbishop O'Brien would do nothing, because his chaplains were telling him it was the command that was causing problems. He would never take my word over that of a priest. The Vatican office responsible for issues involving the traditional Latin Mass was completely uninterested.

The general officers, aside from the courageous Gen. Callender, would do nothing, because the chaplains were forcing them to back off. Senator Lott had been assured by the Military Liaison that there was no need for my request.

Over the next few weeks, I considered legal action. At the recommendation of a local priest (by now we had a close relationship with several wonderful priests in the Brussels area), I initiated instead an Inspector General Complaint, the results of which were unfortunately predictable. But through the Freedom of Information Act, a lawyer helped me to obtain some interesting information. It turns out that, although we were never able to use the NSA Chapel under the "distinctive faith group" category, some other groups were given access to military chapels without any problem. Col. Haberek was the priest who certified Janet L. Cassle as a Distinctive Faith Group Leader for the Wiccan service at Taylor Barracks in Mannheim, Germany. Janet replaced Tammy Stracke, the previously certified Distinctive Faith Group Leader. I was also provided with letters from a mission coordinator of the Temple of Isis, including one ironically sent on the Feast Day of the Immaculate Conception:

> 8 December 2000
>
> Isis Invicta Military Mission
> Temple of Isis (California)
> Mission Coordinator Rev. Rona J. Coomer-Russell
>
> Dear Sir or Madam:
>
> I would like to thank you for taking the time to work with Janet Cassle in her efforts to assist Military Pagans in your area. We hope that she may serve both the military and her local community well.

DEPARTMENT OF THE ARMY
FREEDOM OF INFORMATION AND PRIVACY ACTS OFFICE
7798 CISSNA ROAD
SPRINGFIELD VA 22153-3166

REPLY TO
ATTENTION OF

February 14, 2001

FOIA #01-0433

Mr. Christopher A. Ferrara
President,
American Catholic Lawyers
  Association, Inc.
Suite One, P.O. Box 277
50 S. Franklin Turnpike
Ramsey, NJ 07446

Dear Mr. Ferrara:

This letter responds to your two identical Freedom of Information Act requests dated January 29, 2001, which were addressed to this office. You request all correspondence related to LTC David L. Sonnier, and his request/petition for the offering of the Traditional Latin Mass at NSA Chapel, NATO HQ, Brussels, correspondence related to the request/petition of SFC Vernon Fowler for the use of the NSA Chapel, and the request by a group of "Pagans" for the use of the military chapel at Taylor Barracks in Mannheim.

As a matter under its purview, we have referred your request to the following agency for action and direct response to you.

Headquarters, U.S. Army Europe/7th Army
Office of the Deputy Chief of Staff, Information Management
ATTN: AEAIM-SM-EB
Unit 29351
APO AE 09014
Tel.: 011-49-6221-57-8804

If this office can be of further assistance to you, please call Mrs. Mary Nance Jordan of my staff, or contact me, at the above address. The telephone number is (703) 806-7141/5698.

Sincerely,

Rose Marie Christensen
Chief, Freedom of Information
  and Privacy Acts Office

Printed on ♻ Recycled Paper

We have chosen to assist Janet in her quest to begin religious services IAW AR 165-1 Chaplains Activities in the U.S. Army section 5.5. Our Temple is an IRS tax-exempt religious organization...We were incorporated in the state of California in June of 1996.

Included is Janet's Certificate of Sanction. She is hereby authorized to officiate religious services in accordance with our Military Mission Guidelines. She is given the

American Ca✝holic Lawyers Association, Inc.

50 South Franklin Tpk., Suite 1
PO Box 277
Ramsey, New Jersey 07446

Telephone (201) 236-1799
Facsimile (201) 236-3965

April 5, 2001

Headquarters, U.S. Army Europe/7th Army
Office of the Deputy Chief of Staff
Information Management
ATTN.: AEAIM-SM-EB
Unit 29351
APO AE 09014

      Re: LTC David L. Sonnier
      Freedom of Information Act Request

Dear Sir/Madam:

Please be advised that we represent LTC David L. Sonnier, U.S.A., who has been trying, through the proper military and Catholic Church channels, to establish a regular celebration of the Traditional (Tridentine) Latin Mass for American servicemen in Brussels.

On January 29, 2001, we forwarded a detailed Freedom of Information Act request to the Department of the Army in Springfield, Virginia. Miss Rose Marie Christensen, of the U.S. Army's Springfield headquarters, forwarded a preliminary response to us on February 14 stating that our F.O.I.A. letter had been passed on to you. Copies of both letters are enclosed.

We look forward to the Army's formal response to our F.O.I.A. request on behalf of Lt. Col. Sonnier in the near future.

Very truly yours,

John A. Obriski
Paralegal

JAO:jao
Enclosures
xc: LTC David L. Sonnier

---

authority to perform all duties required of a DFGL. She will be under the direct supervision of Rev. Lisa Mickles, our European secretary.

If you have any questions about the Temple of Isis, Isis Invicta Military Mission you are welcome to contact me at the above addresses or call me. You are also welcome to contact Rev. Mickles.

In Service,

Rev. Rona J. Coomer-Russell
Mission Coordinator

How it was that our simple request could possibly be denied when *this* was allowed by Catholic priests? How could that be? Was this how the "Distinctive Faith Group" clause was to be used? These people had the full support of Fr. Haberek, a Catholic priest. Meanwhile, he had continuously refused to allow priests of the Institute of Christ the King and the Priestly Fraternity of Saint Peter to offer Mass in military chapels.

As I mused over these findings, Gen. Callender walked into my office and closed the door.

"Sir!" I stood.

"Sit down, Dave."

I was not accustomed to having a flag officer walk into my office, so I nervously sat, waiting for the bad news his face foretold.

"Dave, you told me something a few weeks ago that I thought was rather odd at the time, but I want to ask you about it now."

"Yes, sir?"

"You said, at the time that you wrote to Senator Lott, to tell you if there were any repercussions, and that you would let the Senator know."

"Yes, sir, I recall saying that."

"I couldn't believe you told me this. In fact, at the time, I thought that it was inappropriate to suggest such a thing. How could there be any repercussions for just doing my job? But then later, when I was at USAREUR Headquarters, Col. Haberek told me that I would regret having become involved in this issue."

My heart sank, as it dawned on me what he was telling me.

"Well, I just got off the phone with General Officer Management, and it's not looking too good for me," he continued. "Basically, I'm being asked to retire. And I'm being told that the officer behind it, the Chief of Staff of the Air Force, is a staunch Catholic."

"Sir, if he's a 'staunch Catholic,' he has no problem doing as the pope says and allowing the Latin Mass to flourish. Everywhere. Even in the military. If there's someone taking revenge against you for ensuring that we get treated like anyone else serving on active duty, he is certainly not a 'staunch Catholic.' Anyway, you will never regret what you've done, I assure you of that."

It was too much for me to bear. This kind, courageous, decent officer was being asked to retire because he supported traditional Catholics getting the same treatment as any other group of servicemen.

I submitted my retirement paperwork.

CHAPTER 20

FAREWELL

AVING THE PERMISSION OF ARCHBISHOP O'Brien to use the chapel had turned out to be useless. Even so, I submitted a simple request with my notification to His Excellency that I was retiring.

9 May 2001

Most Rev. Edwin F. O'Brien

Your Excellency:

I have made the decision to retire from the U.S. Army this summer. My retirement will be on 29 June 2001 at the NATO Support Activity, and on that day, I would like to have Mass celebrated for my family and friends who will be in attendance at the retirement ceremony. Given that permission has already been given by yourself for the celebration of the Mass according to the classical Roman rite (*Missale Romanum* 1962) I would like to request that this farewell Mass be celebrated according to this rite.

In attendance will be more than one hundred friends, who would greatly appreciate a Latin Mass. I have asked Father William Hudson, a friend of the family, to celebrate the Mass if your permission is given. As he knows Msgr. Marceaux well, I imagine that no opposition would be presented by the Army chaplains. I envision a celebration of the Mass at about 1:30 p.m. with the retirement ceremony at 3:00 p.m.

I would like to take this opportunity of thanking you for the permission that you granted regarding the celebration of the Latin liturgy. I deeply regret the circumstances that proved an obstacle to this permission being applied. Be assured of my prayers and devotion.

Very Respectfully Yours in Christ,
David L. Sonnier
Cc: Commission *Ecclesia Dei*

Lorri and I had become friends with many of the Belgian Catholics while living there. Most of the families at *L'Institut Saints Pierre et Paul*, where our oldest three children went to school, were dedicated to the traditional Mass, and they were all invited to my retirement ceremony. We had met numerous Catholics at a small parish church we attended on Sundays in Cortil-Noirmont, a remote location with a particularly good priest. We had made new friends on an annual pilgrimage that started in Dinant and ended about ten kilometers away at Notre Dame de Foy, a church built in 1623 that had become a pilgrimage site.

Our children attended L'Institut Saints Pierre et Paul,
where they received a genuine Catholic education.

Over the course of the three years that we were in Belgium,
we participated in this pilgrimage every year.

Michael Davies can (just barely) be seen, holding the Welsh flag.

The SSPX was renovating an old church in the downtown area of Brussels as we were preparing to leave Belgium. Saint Joseph's was an Italian-Renaissance-style church built between 1842 and 1849, in the Leopold district. From the time it was consecrated in 1849 until sometime in the 1980s it was entrusted to the Redemptorists. As religious life and Mass attendance plummeted after the Second Vatican Council, the Syrian Orthodox took it over for a few years. In 2001 the SSPX bought it.

Sometime before departing the country we heard about their first Mass and we made it a point to attend. We had never attended Mass with the SSPX until now. As we entered the packed church and looked around, to my surprise there were dozens of people that we knew. Apparently, attendance at Mass with the SSPX was something that many of our Belgian Catholic friends did but nobody discussed. People that I knew from our choir at Cortil-Noirmont were participating in the choir, which sang a beautiful polyphonic Mass. People we knew from our children's school, *L'Institut Saints Pierre et Paul*, were seated next to us. The SSPX bishops would not have their excommunication lifted for a few more years, but it was then and there that I began to fully comprehended the extent of the injustice that had been done. This was where the large families were. This was the vibrant, youthful crowd that would be the future of the Church. These Catholics had managed to aquire a beautiful old church and preserve the traditional Catholic liturgy, doctrine, and prayer life without the blessing of the hierarchy, but they were supposedly the culprits. Meanwhile, so many other magnificent churches and cathedrals throughout Europe sat empty because the liturgy and spirituality that they had been built for was now forbidden.

CHAPTER 21

INTROIBO AD ALTARE DEI

O N JUNE 29, 2001, ON THE DAY OF MY RETIRE-
ment from the U.S. Army, we finally succeeded. Finally,
after six years, twenty-three meetings with general
officers, more than seventy-five letters, at least ten of them to the
Vatican, a personal trip to the Vatican, an appeal to the Senate
Majority Leader, a media news break, and an Inspector General
complaint, we were able to use the NSA Chapel for a Latin Mass
according to the 1962 Missal one hour before my retirement.
Both the NSA Chaplain and the 80th ASG Chaplain were new,
and neither of them seemed willing to offer any resistance.

Retirement ceremony, with Lieutenant General David Weisman,
at the NATO Support Activity, June 29, 2001. Earlier in the day we
finally succeeded in having a Traditional Latin Mass just a short
distance from where this ceremony took place.

I scheduled the chapel just like any ordinary, non-Catholic serviceman would—just as we should have been able to do all along. About twenty people happened to find out about the hastily planned Mass. Knowing the infinite value of just one Mass, I was able to breathe a sigh of relief. While this long saga might look like a failure, it was clear that God had answered our prayers, even if at the last minute, and that our labors and efforts on behalf of what Fr. Faber had once referred to as "the most beautiful thing this side of heaven" were not in vain. Our little group prayed the same prayers that so many had prayed before us: St. Edmund Campion, St. Thérèse of Lisieux, St. John Vianney, St. Joan of Arc, and all the other saints who labored to build a Christian civilization in their time. We prayed the same prayers that were prayed by all five Catholic chaplains who received the Medal of Honor in the service of the U.S. Armed Forces: Fr. Emil Kapaun, Fr. Vincent Capodanno, Fr. Joseph O'Callahan, Fr. Charles Watters, and Fr. Angelo Liteky. The same prayers that all five of my father's brothers had prayed whenever there was a Catholic chaplain available during World War II.

Finally, those dangerous and forbidden words, such a threat to our unbelieving age, were uttered in prayer within a United States Military Chapel on the Feast Day of Saints Peter and Paul:

**Introíbo ad altáre Dei,
ad Deum qui laetíficat
juventútem meam.**

*I will go unto the altar of
God, unto God who
giveth joy to
my youth.*

EPILOGUE

AT THE TIME THAT THIS SEQUENCE OF EVENTS was recorded, it was not unusual to have a request for the traditional Latin Mass treated with disrespect and contempt, regardless of how politely the request was made. In the first edition of this book, I challenged anyone who doubted this assertion to attempt to have Gregorian chant included in the Mass on the following Sunday, or next month, or any time in the future at their home church. I challenged the reader to request of his or her priest authorization to organize a schola, so that the *Kyrie Eleison, Gloria, Credo, Sanctus*, and *Agnus Dei* could be given "pride of place" in the liturgy, in accordance with the Second Vatican Council's Constitution on the Sacred Liturgy (no. 116), or so many of the other documents on the liturgy that have come along since. At that time the petitioner would have typically been given a multitude of reasons why it was not possible. If nothing else, the response would be "Latin is not the norm," and "we've become accustomed to it this way." The problem was often a basic lack of charity, but sometimes it was confusion over the status of the traditional Latin Mass and a broader misunderstanding of what was actually stated in the Second Vatican Council's documents. Many of the priests and bishops I was dealing with seemed to have never heard anything about the 1988 motu proprio *Ecclesia Dei*.

Fr. Eugene Dougherty composed the following poem for *Cristian Order*, a Catholic periodical published in the UK.[1] Although I am aware of no scientific study to determine how many requests were turned down by bishops, it is quite certain that a negative response was the norm. The poem is reprinted with the permission of Fr. Dougherty.

The Bishop's Reply Song

> I found your request in the mail yesterday
> And hasten to answer without a delay,
> To settle your problem and issue a writ
> Allowing whatever the Canons permit.

---

[1] It appeared in the August/September 2000 issue.

A delicate question you boldly propose
In language intemperate that sounds bellicose.
Ignoring the precepts of Vatican II
You lecture your bishop on what he must do.

We bishops already in solemn conclave
Agreed on a uniform way to behave:
Whenever requested to say the Old Mass,
Ignore the petition and soon it will pass.

For those who request it are often quite old
And suitably brainwashed to do as they're told.
They open their wallets and willingly pay
Since they are conditioned to humbly obey.

But like a good shepherd I offer this boon...
If you can arrange for a Mass on the moon.
Thus, granted you're willing and rightly disposed,
With kindest regards, there's a ticket enclosed:

"Being over 65 and having sworn to the validity of the
Novus Ordo Mass, the bearer Homer Simplex is entitled
to attend one (1) Tridentine Mass on the Moon at any
church or semi-public oratory where available on the
second Sunday of any month beginning with the letter
'm' at 2 o'clock L.T. (Lunar Time), weather permitting.
Nontransferable, non-refundable."

Much has changed since then.

In 2007, Pope Benedict XVI clarified the issue of the 1962 *Missale Romanum* with the motu proprio *Summorum Pontificum*,[2] which gave priests anywhere the right to offer the traditional Mass and gave petitioners a right to appeal to the Holy See when their petitions were being denied. Most of the U.S. bishops seem to have welcomed this long-overdue clarification and used it to the benefit of the faithful. Soon there were many flourishing Latin Mass communities throughout the U.S., including one at Fort Hood, Texas. This community was established in 2015, but it came to an unfortunate end only two years later when the priest who was the chaplain of the community retired.[3] A founding member of the Fort Hood Latin Mass community,

---

[2] *Summorum Pontificum* was issued in 2007, shortly after the publication of the previous edition of this book, *Rightful Aspirations.*

[3] Brian Williams, "Archdiocese of Military: Better to Have No Priests than Traditional Ones," *Liturgy Guy*, June 17, 2017, https://liturgyguy.com/2017/06/17/archdiocese-of-military-better-to-have-no-priests-than-traditional-ones/.

Sergeant Major John Proctor, who has since retired but was at that time serving as the Third Armored Corps Chaplain Sergeant Major, wrote to Archbishop Timothy Broglio of the Archdiocese for the Military Services (AMS) to seek help in finding a solution that would enable the community to continue. He also offered a suggestion to the AMS for the larger problem of the shortage of Catholic chaplains. Archbishop Broglio had drawn attention to this shortage in 2015, when he informed his brother bishops at the annual USCCB meeting that only eight percent of the chaplains on active duty were Catholic. The suggestion that SGM Proctor offered, in order to deal with the immediate problem of the Fort Hood Latin Mass community and the larger problem of the shortage of Catholic chaplains, was to invite "traditional priests," those offering Mass according to the 1962 liturgical books or earlier, to consider joining the military as chaplains. In a letter to Archbishop Broglio, SGM Proctor said:

> According to the Center for Applied Research in the Apostolate (CARA), the average age of a priest in the U.S. is 64. In the Priestly Fraternity of St. Peter (FSSP), it is 37. The FSSP and the Institute of Christ the King Sovereign Priest (ICKSP) have almost 500 priests with over a hundred in formation now. The Priestly Society of St. Pius X (SSPX) may soon be regularized by the Holy Father and granted a personal prelature. They have 600 priests and over 200 seminarians in formation now.... Priest-chaplains for the future may be available from these traditional priestly societies if we actively recruit them and permit them to serve in their charism of exclusively using the 1962 liturgical books.

Archbishop Broglio's response came as no surprise. It reflected the same thinking that had existed twenty years earlier. He believed, from a pastoral point of view, that he would be doing a disservice to his flock if he did not provide the liturgy to which most Catholics were accustomed. He would not invite a traditional priest into the chaplaincy, because there may be some people on the base who would not have their liturgical expectations met by such a priest, whose prayers at the Mass would be in Latin. He equated the situation to that of an Eastern-rite Catholic priest serving as a chaplain. Setting aside the flaws in that comparison, from the point of view of Archbishop Broglio

the two situations were similar because his desire was to meet
the liturgical expectations of most Catholics. If this was, and
remains, the goal, we must question whether it is realistic. If
those Catholic communities in the U.S. that are emotionally or
spiritually bound to the revised Missal are not providing sufficient priests to serve as military chaplains, perhaps it is time
for the AMS to relax the goal of meeting the liturgical expectations of that same population within the U.S. Armed Forces.
There are plenty of servicemen who would welcome the addition
of priests who are exclusively of the traditional Latin rite and
priests coming from Eastern rites, and who would prefer having
such priests to not having any priest at all.

The inclusion of traditional priests in the chaplaincy (those
coming from *Ecclesia Dei* orders, for example) would begin correcting the record regarding some misunderstandings that exist
among U.S. Catholics. It is time to correct false narratives that
proclaim that Latin, the language of the Roman rite, is somehow
forbidden and that Gregorian chant, which is supposed to have
"pride of place" in the liturgy, is somehow a thing of the past.
To be able to play a role in correcting these prolonged misunderstandings of the Second Vatican Council should be sufficient
incentive for the inclusion of the traditional Latin Mass wherever and whenever possible on military installations. Hopefully,
enough time has passed that those Catholic chaplains who are
bound to the false narratives about Vatican II have moved on
and will no longer present obstacles. It is time to make their
sad legacy a thing of the past.

As of 2014, Fort Hood had more than 45,000 assigned soldiers
and nearly 9,000 civilian employees. Within a population this
size, and on another base with an even greater population such
as Fort Bragg, North Carolina, the problem of "liturgical expectations" is not a problem at all. The Latin Mass community at
Fort Hood, despite being less than two years old when it ended,
had 120 members and an average weekly attendance of 45–55
people. The question should not be whether or not the liturgical expectations of the faithful are met, but whether or not the
sacraments are available. Were a traditional priest assigned
to one of these larger installations in the U.S. or abroad, the
liturgical expectations of some people would be met, but more
importantly, the sacraments would be available.

In a 2025 article, Fr. Paul Hurley, former Chief of Chaplains of the U.S. Army, was interviewed along with two Catholic chaplains from Fort Bragg concerning the current shortage of Catholic chaplains. Fr. Hurley correctly stated the problem:

> The nature of Catholic worship means there's always an urgent need, Father Hurley said. "Catholics need priests. Their sacramental life is not like other denominations," which can hold services without clergy present. But for the faithful, "there is no other way to fulfill the Catholic requirements."[4]

One of the chaplains from Fort Bragg who was interviewed stated that, given that Fort Bragg now has about 61,000 military personnel and about 14,000 civilian employees, his estimate was that he had about 12,500 parishioners. Within such a population of Catholics there are many who would appreciate the opportunity to attend the traditional Latin Mass. If the estimate of 7.1% is correct,[5] out of a population of 12,500, about 887 of those Catholics would attend the Latin Mass. Should the liturgical sensibilities of these Catholics simply be ignored?

As for the traditional priest assigned to smaller bases, of which there are hundreds worldwide, perhaps the time has come for the Archdiocese for the Military Services to help servicemen to understand the diversity and richness of our Catholic tradition. Priests who offer the sacraments according to the 1962 *Missale Romanum* are very adept at ensuring that visitors understand the way prayers are being offered and the proper position and disposition for receiving Holy Communion. Assuming that such a priest is the only Catholic chaplain on a small installation, as opposed to a mega-installation such as Fort Bragg or Fort Hood, what harm could there possibly be? If the shortage of Catholic chaplains is genuinely serious, isn't this better than the alternative? Is it better to have *no* priest than to have a *traditional* priest?

---

4 Kurt Jensen, "Need for more Catholic Army chaplains to serve military flock as great as ever, say two priests," *OSV News*, June 20, 2025, https://catholicreview.org/need-for-more-catholic-army-chaplains-to-serve-military-flock-as-great-as-ever-say-two-priests/.

5 "13% of all US Catholics attended a Traditional Latin Mass in the Past 5 Years: Traditional Catholics are 7.1% of all practicing Catholics in America," *Rorate Caeli*, June 26, 2025, https://rorate-caeli.blogspot.com/2025/06/13-of-all-us-catholics-attended.html.

Since Archbishop Broglio compared the priest who offers Mass exclusively in a pre-conciliar rite to an Eastern-rite Catholic priest, this raises other questions. Is there really a need to require Eastern-rite priests to be biritual to serve as chaplains? What could possibly be wrong with allowing them to offer Divine Liturgy in their own rite, even when those attending are not familiar with it? My family derived tremendous spiritual benefit from spending time with the Maronite Catholics in Fayetteville, NC. I often invited fellow servicemen to attend with us, and it was always an eye-opening experience to come to understand that this other tradition was also a part of the universal Church. Many of our servicemen spend time living and working in parts of the world where one of these other rites may be the norm. Rather than excluding such rites altogether, why not lay out procedures for having a chaplain of a different rite briefly explain, at the beginning of each service, the proper manner and disposition for receiving the Eucharist, and how the liturgy that they are attending may differ from what they are likely accustomed to? Of all possible congregations, a congregation of military people is the least likely to object to attending Mass or Divine Liturgy in a rite with which they are not familiar.

The AMS currently has guidelines for "Sundays/Holy Day Celebrations at a chapel in the absence of a priest," found nested in their web page.[6] These guidelines include:

> 2.2. If a Catholic priest is not available to celebrate a Mass, the faithful should be directed to celebrate at a civilian Catholic parish.

> 2.3. In the absence of a Catholic priest, Catholics can observe the sacredness of Sundays by participating in a Catholic liturgy led by a deacon or authorized EMHC.[7] The Archbishop for the Military Services, through his delegate, the Vicar General or Chancellor, must explicitly give approval for this rite.

> 2.4. The deacon or EMHC must clearly specify that the Sunday Celebration in the Absence of a Priest is not a Sunday Mass, explaining that the rite is a temporary response and not considered an ideal solution to the present situation.

---

[6] https://files.milarch.org/clergy/norms-for-emhc-7jan14.pdf.

[7] Extraordinary Minister of Holy Communion.

It would seem reasonable that if approval for a rite that *doesn't* include a priest can be given, per paragraph 2.3, then approval for an Eastern or traditional Latin rite that *does* include a priest could also be given. The requirement for providing an explanation in paragraph 2.4 above is no less burdensome than an explanation describing how the Latin Mass or Divine Liturgy differs from that to which some of those attending may be accustomed. I am aware of at least two occasions, in the diocese where I currently reside, in which an FSSP priest substituted for a diocesan priest who could not otherwise find a replacement. On both occasions Mass was preceded by a simple explanation of the manner for receiving Holy Communion and that the Mass would be in Latin. The world did not end.

The Archdiocese for the Military Services follows the model of "recruiting" priests by borrowing them from dioceses that are willing to loan them. Apparently they intend to continue to do so. Beggars cannot be choosy. If they continue with this model instead of putting forth the resources and effort to build their own seminary, they should accept the services of any priest, regardless of rite.

If the Archdiocese for the Military Services cannot lift the requirement that Eastern-rite priests be biritual to serve as chaplains, perhaps it's time to end the AMS monopoly on the Catholic chaplaincy. In fact, why is the AMS the only endorsing agency for Catholic chaplains? The list of endorsement agencies[8] includes many Protestant denominations and Orthodox, Jewish, and other religious groups. For Catholics, the only endorsement agency is "Roman Catholic Church," led by Archbishop Broglio. Why is that? Perhaps, given the inability of the Archdiocese for the Military Services to provide sufficient priests for our service personnel, there should be other Catholic endorsing agencies that are not specifically "Roman Catholic." What could be wrong with having a "Byzantine Catholic" endorsement agency or a "Traditional Roman Catholic" endorsing agency, for example? The Archdiocese for the Military Services has only existed as an independent archdiocese since 1984, by the decree *Apostolica Sedes*. This decree states nothing about rites, other than that the "faithful in the Latin and Eastern Churches" affiliated with

---

8 https://prhome.war.gov/M-RA/MPP/AFCB/Endorsements/.

the military are members of the Military Vicariate. Where did
the idea originate that an Eastern-rite priest must offer Mass
in the "Latin rite"—a "Latin rite" that, moreover, excludes Latin
more often than not, and includes an abundance of problematic
liturgical anomalies described in this book?

If the idea of alternative endorsing agencies is too difficult
to consider, then what would be wrong with having, among
auxiliary bishops, one who comes from an Eastern rite? In the
"Statutes of the Archdiocese for the Military Services, USA,"[9] we
find the following:

> A suitable number of Auxiliary Bishops shall be appointed,
> at the pleasure of the Apostolic See, to assist the Mili-
> tary Ordinary in his pastoral ministry and in the care
> of souls in this far-extended jurisdiction. The Auxiliary
> Bishops shall be appointed according to the pastoral
> necessity of the Archdiocese for the Military Services,
> U.S.A. according to the norm of c. 403 §1. VII.

In July 2021, Pope Francis felt the need to restrict what
appeared to be the ongoing restoration of preconciliar rites.
He did this through *Traditionis Custodes*, which at the time of
this Epilogue's writing is still in effect. This document should be
summarily dismissed by any rational Catholic for any number
of reasons, not the least of which is that it is based on a lie.[10]
Even if *Traditionis Custodes* is to be taken into consideration,
one of its features is that the 1962 Missal is not to be used in
parish churches. The Archdiocese for the Military Services has
chapels, not parish churches.[11] The restriction against using
the 1962 Missal in parish churches, therefore, should not be
an obstacle to allowing servicemen access to the 1962 Missal
in chapels. *Traditionis Custodes* declares that the post-conciliar
liturgical books are to be the unique expression of the Roman

---

[9] www.milarch.org/ams-priests-manual/statutes/.

[10] See "Official Vatican Report Exposes Major Cracks in Foundation of *Traditionis
Custodes,*" Diane Montagna's Substack, July 1, 2025. For discussion of the document's canonical flaws, see Réginald-Marie Rivoire, *Does "Traditionis Custodes"
Pass the Juridical Rationality Test?* (Os Justi Press, 2022). For a discussion of its
theological and pastoral flaws, see Peter Kwasniewski, ed., *From Benedict's Peace
to Francis's War: Catholics Respond to the Motu Proprio* Traditionis Custodes *on
the Latin Mass* (Angelico Press, 2021).

[11] Chapels, oratories, and shrines are all examples of non-parish locations where
Mass may be offered even according to the strictures of *Traditionis Custodes*.

rite. However, *Traditionis Custodes* also gives the diocesan bishop "exclusive competence" to authorize the use of the 1962 Roman Missal in his diocese. It is difficult to see how the Archdiocese for the Military Services, which would have competence to authorize the 1962 *Missale Romanum*, could have any objections to simply authorizing it in the case of a chaplain coming from a traditional order whose members already enjoy the normative use of the old liturgical books according to the provisions of their Vatican-approved constitutions. Likewise for the liturgy used by the Anglican Ordinariate, now commonly referred to as "Divine Worship."

The recommendations thus far have focused on the liturgy, the Mass, and the urgent need to lift restrictions on traditional forms of Catholic worship of both the East and the West. The primary reason we have chaplains in the armed forces is for battlefield ministrations. As important as the Mass is, Catholic chaplains are irreplaceable on the battlefield to offer baptism, confession, and extreme unction. Unlike with the Mass, these sacraments, even in their traditional form, can be administered in the vernacular, except for prayers of exorcism and absolution. Last rites are sometimes administered to those who may not even be conscious. To restrict traditional priests from the military chaplaincy is to potentially deny battlefield sacraments to injured or dying servicemembers. An injured or dying soldier would not complain about Latin or any other language used to apply sanctifying rites.

Finally, there is something even more important at stake, and that is the loss of a traditional Catholic understanding of morality in the declaration and conduct of war and the use of deadly force. The contemporary understanding of moral conduct in the profession of arms is at an all-time low. A young Catholic officer recently pointed out to me that nobody in his chain of command seemed to know anything at all about the just war criteria laid out by St. Thomas Aquinas, and in some cases, they have never even heard of it. There are, in fact, very strict limitations on the use of deadly force that go beyond the laws of warfare laid out by the Geneva Conventions, UCMJ, and other U.S. laws. Those restrictions should be well known and understood from the highest level of national command authority down to the platoon level. Instead we find that deadly

force is thrown around at the whim of whoever happens to be
president at the time, with no apparent consideration for the
loss of human life. There is no way that this would be supported
or even tolerated by our armed forces if the ranks were full of
tradition-minded Catholics having a proper understanding of
the just war guidance given to us by St. Thomas.

Where, among the senior officers, do we find any Catholics
voicing objection to the improper use of military force? One of
the unfortunate results of this free-for-all prevailing in the post-
conciliar era is the widespread notion that rules are optional. The
blatant disregard of liturgical laws, and the inclusion of practices
condemned in the strongest terms throughout the history of the
Church, including in 1970, 1980, and 2004,[12] are just a symptom
of the ongoing post-conciliar corrosion of Catholic influence
within our military. This book described a pattern of the type
of misbehavior that further corrodes the desperately needed
Catholic influence on our use of military force. The behavior of
the senior chaplains and the AMS during the period covered
in this book was somewhat like that of petulant children who,
upon discovering that there were no consequences for their
disobedience to the pope, did what they wanted. How can the
officers and senior NCO leadership of a large and dangerous
military be spiritually guided by priests who don't honor the
spiritual authority over them? If these priests disregard the
guidance of the Church in what they wrongly consider to be
"little things," such as liturgy, will they be up to the task of providing spiritual guidance in questions of life and death in the
use of deadly force? *Lex orandi, lex credendi, lex vivendi* — how
we pray shows and shapes what we believe and the way we live.
The doctrinal and liturgical traditions of the Church come as
a package deal. Discarding the traditional liturgy has brought
America to a dark place, where the most monstrous behavior
and barbaric acts are tolerated by our military because the
Catholic understanding of just war and the value of human life
have become collateral damage.

The contempt and disregard that the AMS had for *Ecclesia
Dei* (1988) as described in this book was followed by a similar

---

[12] *Liturgicae Instaurationes* (1970), *Inaestimabile Donum* (1980), and *Redemptionis
Sacramentum* (2004).

disregard for *Summorum Pontificum* (2007). It is absolutely time to end this kind of thinking, and time for the AMS to understand that a greater respect for Catholic tradition is demanded by circumstances. The tradition of the just war has been nearly completely lost through the past decades of post-conciliar chaos. The U.S. military can be influenced in the direction of a greater respect for human life if the AMS will take the simple step of inviting traditional priests into the Catholic chaplaincy and promoting a traditional form of worship. Such a step would undoubtedly end the lamentable shortage of Catholic chaplains. The spiritual health of our service personnel is at least as important as their physical health. We cannot continue to deprive them of the ministrations of excellent traditional Catholic priests out of slavish obedience to an agenda from the 1970s.

I hope and pray that those able to remedy this situation will be emboldened to fight the good fight.

www.ingramcontent.com/pod-product-compliance
Lightning Source LLC
Chambersburg PA
CBHW032227050726
47591CB00001B/299